A TEXT BOOK OF

COMPARATIVE ANATOMY OF VERTEBRATES

[Paper III]

FOR
B.SC. PART - I (ZOOLOGY) : SEMESTER - II
[CREDITS 2]

**As Per New Choice Based Credit System (CBCS Pattern) Syllabus of
Punyashlok Ahilyadevi Holkar Solapur University, Solapur w.e.f. June 2019**

DR. KISHORE R. PAWAR
M.Sc., Ph.D.
Former Principal,
Karmveer Abasaheb Alias
N. M. Sonawane Arts, Science
and Commerce College,
Satana, Dist. Nashik.

DR. ASHOK E. DESAI
M.Sc., Ph.D.
Former Associate Professor,
P. G. Department of Zoology,
KTHM College,
Nashik - 422002.

DR. DAMA L. B.
M.Sc., Ph.D.
D.B.F. Dayanand College
of Arts and Science, Solapur District - Solapur

N5214

COMPARATIVE ANATOMY OF VERTEBRATES   **ISBN 978-93-89825-36-7**

First Edition : **January, 2020**

© : **Authors**

Published By:

NIRALI PRAKASHAN
Abhyudaya Pragati, 1312, Shivaji Nagar
Off J.M. Road, PUNE – 411005
Tel - (020) 25512336/37/39, Fax - (020) 25511379
Email : niralipune@pragationline.com

➤ **DISTRIBUTION CENTRES**

PUNE

Nirali Prakashan : 119, Budhwar Peth, Jogeshwari Mandir Lane,
Pune 411002, Maharashtra
Tel : (020) 2445 2044, 66022708, Fax : (020) 2445 1538
Email : bookorder@pragationline.com,
niralilocal@pragationline.com

Nirali Prakashan : S. No. 28/27, Dhyari, Near Pari Company, Pune 411041
Tel : (020) 24690204 Fax : (020) 24690316
Email : dhyari@pragationline.com,
bookorder@pragationline.com

MUMBAI

Nirali Prakashan : 385, S.V.P. Road, Rasdhara Co-op. Hsg. Society Ltd.,
Girgaum, Mumbai 400004, Maharashtra
Tel : (022) 2385 6339 / 2386 9976, Fax : (022) 2386 9976
Email : niralimumbai@pragationline.com

➤ **DISTRIBUTION BRANCHES**

JALGAON

Nirali Prakashan : 34, V. V. Golani Market, Navi Peth, Jalgaon 425001,
Maharashtra, Tel : (0257) 222 0395, Mob : 94234 91860

KOLHAPUR

Nirali Prakashan : New Mahadvar Road, Kedar Plaza, 1st Floor Opp. IDBI Bank
Kolhapur 416 012, Maharashtra. Mob : 9850046155

NAGPUR

Nirali Prakashan : Above Maratha Mandir, Shop No. 3, First Floor,
Rani Jhanshi Square, Sitabuldi, Nagpur 440012, Maharashtra
Tel : (0712) 254 7129; Email: niralinagpur@pragationline.com

DELHI

Nirali Prakashan : 4593/15, Basement, Agarwal Lane, Ansari Road, Daryaganj
Near Times of India Building, New Delhi 110002
Mob : 08505972553

BENGALURU

Nirali Prakashan : Maitri Ground Floor, Jaya Apartments, No. 99, 6th Cross,
6th Main, Malleswaram, Bangaluru 560 003, Karnataka
Mob : +91 9449043034
Email: niralibangalore@pragationline.com

www.pragationline.com   **info@pragationline.com**

This Book is Dedicated to my Respected teacher and great Zoologist

Prof. P. R. Gadre of Pratap College, Amalner Dist. Jalgaon.

Prin. Dr. Kishore Pawar

PREFACE

The authors are indeed very happy to present this book **'Comparative Anatomy of Vertebrates'** for the students of B.Sc. Part I Zoology, Semester II of Punyashlok Ahilyadevi Holkar Solapur University, Solapur.

The book has been written according to the new revised CBCS syllabus.

There was a long felt need of the students as well as teachers community for a text book which covers the entire syllabus prescribed by Board of studies. The present book is an outcome of our sincere efforts. We tried our level best to present the subject matter in easy style and in a comprehensive manner. The text book is profusely illustrated with number of clear line drawings.

No doubt, there are several textbooks written by Indian and foreign authors on the subject, but they are costly and number of copies are very limited in the college libraries. The students can not get the matter on prescribed syllabus in one book and they also can not afford the costly books. Therefore, we have presented all the topics in one book in a low price. We sincerely feel that this book will fulfill the requirements of students and teachers.

We are thankful to Shri. Dineshbhai Furia, Shri. Jignesh Furia, Mr. Malik Shaikh, Mrs. Roshan Khan, Mrs. Anjali Muley and the entire staff of Nirali Prakashan for taking keen interest in publishing this book and bringing out in time.

Constructive suggestions for improvement of the book are most welcome.

– Authors

SYLLABUS

Unit 1. Integumentary System (4)

Derivatives of Integument with reference to Glands and Digital Tips

Unit 2. Skeletal System (4)

Appendicular and Axial Skeleton in Vertebrates

Unit 3. Digestive System (5)

Brief Account of Alimentary Canal and Digestive Glands

Unit 4. Respiratory System (5)

Brief Account of Skin, Gills, Lungs, Air Sacs and swim Bladder

Unit 5. Circulatory System (4)

Evolution of Heart and Aortic Arches

Unit 6. Urinogenital System (4)

Succession of Kidney, Evolution of Urinogenital Ducts

Unit 7. Nervous System (4)

Comparative Account of Brain

***

CONTENTS

***

Chapter **1** ...

Integumentary System

Contents ...

1.1 Derivatives of Integument

(A) Epidermal Glands:

The epidermis of the integument also forms different types of glands called epidermal glands. These glands are give rise from malpighian layer of epidermis. According to their structure they are unicellular, tubular and alveolar glands. They may be multi-cellular, compound or branched. These glands are lined by cuboidal or columnar epithelium.

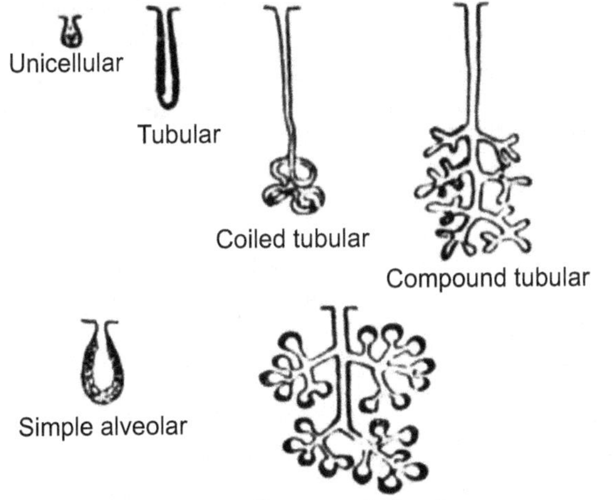

Fig. 1.1: Epidermal glands of unicellular, tubular and alveolar types

Depending on the function of glands, they are named. There are major types of epidermal glands namely, mucous glands, poison glands, oil glands, wax glands, sweat glands, sebaceous glands, uropygia glands, scent glands, and mammary glands.

(1) **Mucous glands:** They may be unicellular or multicelluar, Among the unicellular glands, are mucous gland cells, granular cells and breaker cells. These glands secrete mucin which keeps the skin moist and slippery. Mucous also protects the skin against bacteria and fungi. They extend from malpighian layer to the surface of skin. These glands are abundant in fishes and amphibian skin.

(2) **Poison glands:** In case of certain fishes and amphibians like toads, poison glands are present in the skin. They are actually modified multicellular cutanous glands larger but they are few in number as compared to mucous glands. In case of toad behind the head **parotid glands** are present, which are the aggregation of poison glands. These glands secrete bitter, irritating and toxic secretion which is harmful to the predator.

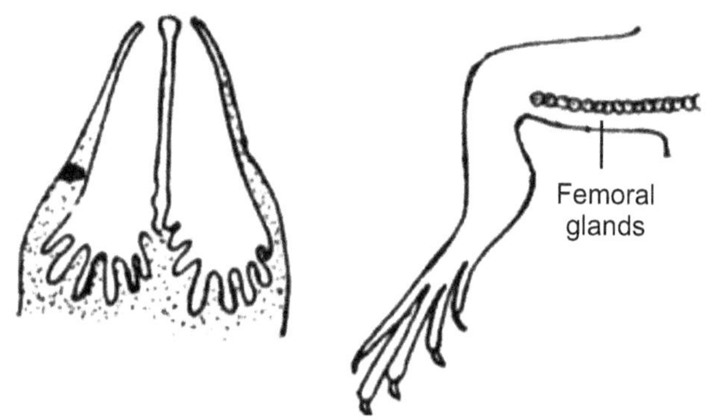

(a) Femoral glands of lizard   (b) Uropygial gland of bird
Fig. 1.2: Epidermal glands

(3) **Uropygial glands or oil glands:** These glands are found in birds. These glands form a prominent swelling just above the tail or uropygium. These are branched and alveolar glands and secrete an oily secretion used for lubricating beak, preening feathers and attracting the opposite sex.

(4) Luminescent glands: There are also called photophores because the glands in deep fishes (e.g. teleosts) serve as light emitting organs. The superficial cells layer forms a magnifying lens, and the basal cells called *luminous* cells and these cells are surrounded below by reflecting pigment cells. They emit very weak light and this light is useful to attract the prey.

(5) Femoral glands: These glands are found in uromastix male lizards on the ventral surface of each thigh. These glands are arranged in a single row from knee to cloacal opening. They secrete a sticky substance which becomes hard when come in contact with air. The secretion forms small spines which are useful to hold the female during copulation.

(6) Sweat glands: The mammals show the large number of sweat glands in the skin. They are slender, coiled tubes embeded in the dermis and they open on the skin surface with their long ducts. These glands secrete some salts and urea dissolved in water. The sweat is also useful in cooling body temperature in mammals in hot environment. Sweat glands are absent in some mammals like scaly ant eaters, whales and dophins. In some mammals, these glands also occur only in particular body parts.

(7) Sebaceous glands: These glands also occur in mammals which are branched, alveolar glands. These glands open into hair follicles. They also open directly on to skin surface such as around the genital organs; tip of nose or edges of lips. These glands secrete oily secretion or sebum (i.e. grease) which keeps the skin and hairs soft, oily, water proof and glistening. The animals such as marine mammals like whales and dophins show absence of hairs hence sebaceous glands are also absent. These glands are also absent in pangolins.

(8) Wax glands: These are called ceruminous glands, present in the external ear canal of mammals. These glands are modifed sebaceous glands. They are also called wax glands because they secrete waxy or greasy secretion called **cerumen**. This wax help in trapping dust particles and small insects in external ear. The meibomian glands are located in eyelids, which spread their oily secretion over the exposed surface of eyeball. Thus, the movements of eyelids is facilitated. These are also modified sebaceous glands.

(9) **Scent glands:** Scent glands are modified sebaceous or sweat glands of mammals. These glands secrete scented secretion which serve to repel the toes or attract the member of opposite sex. The scent glands are located at different places or parts in mammals. They occur between the toes on feet in horse, rhino and goat. In deer, they occur near eyes on head. In case of musk deer these glands occur in naval on abdomen. In the carnivore animals like lion, tiger and rodent, these glands occur around the anus. In zoos and animal houses, foul odour occurs not due to unhygienic conditions but caused by the scent glands of these caged mammals.

(10) **Mammary glands:** Presence of mammary glands is the characteristic of mammals. These are compound tubular glands which produce milk during lactation period for feeding the young ones. They are found not only in females but they also present in male monotremes, primates and other mammals. They do not show nipples or teats and are very similar to sweat glands. In other mammals they possess nipples and are modified sebaceous glands. Distribution of mammary glands and nipples vary with the species. In monotremes the tubular mammary glands have no nipples, they open directly on the surface of a depressed area of the skin, and the young, lick milk from these hairy areas. In *metatherians* and some *entherians* (carnivores, man), the mammary glands open by their ducts on the tip of a nipple which is a raised area of the skin. The breasts of mammal consist of mammary glands and their surrounding

connective tissue and fat. In ungulates the ducts of mammary glands open into a chamber or cistern at the base of a false nipple or teat which becomes elongated, milk is carried from the cistern to the tip of the teat by a secondary tube or duct. In primitive mammals, the mammary glands were arranged along a ventral milk line on each side from the chest to the groin. In higher mammals, mammary glands are thoracic, restricted to pectoral region (bat, elephants, primates) or to an inguinal region (ungulates, whales). Lemurs have a single pair in the arm pits (axillary), but in many mammary glands mammal are axillary, thoracic, abdominal, and inguinal along the milk lines.

1.2 Derivatives of Integument : Digital Tips

Both layers of skin i.e. epidermis and dermis have given rise to various types of derivatives. The epidermis gives rise to integumentary glands, epidermal scales, scutes, beaks, horns, claws, nails and hoofs, and different corneal structures, feathers and hairs. The dermis also gives rise to dermal derivatives such as dermal scales, plates or scutes, fin rays and antlers.

1.2.1 Epidermal Derivatives

(i) **Epidermal scales and scutes:** In the cells of epidermis hard and horny structures develop by the accumulation of scleroprotein called Keratin. These cells are called cornified or keratinised cells and they become dead. The layer of epidermis called stratum corneum cells is keratinised and form hard horny protective exoskeleton in the form of scales, beak, horns, claws, nails, hoofs, feathers, hairs etc. in different vertebrates.

Reptiles show the horny scales on the skin which protect the body from prevention of water loss through the skin surface. In reptiles e.g. lizard scales are thin, small overlapping. The scales on the ventral surface are plateline and help in locomotion are called scutes. The snakes and some lizards show shedding of skin period in one piece called *ecdysis* or *moulting.*

The turtles, tortoises and crocodiles show large thick, rectangular scutes. They are not showing overlapping arrangement but touching each other and supported beneath by dermal bones. The scutes are also slonghed or shed in patches at intervals.

There are also certain modifications of stratum corneum in reptiles occur such as toothless horny beak in turtles, the rattle at the end of the rattle snake and horns in horn toad (a lizard).

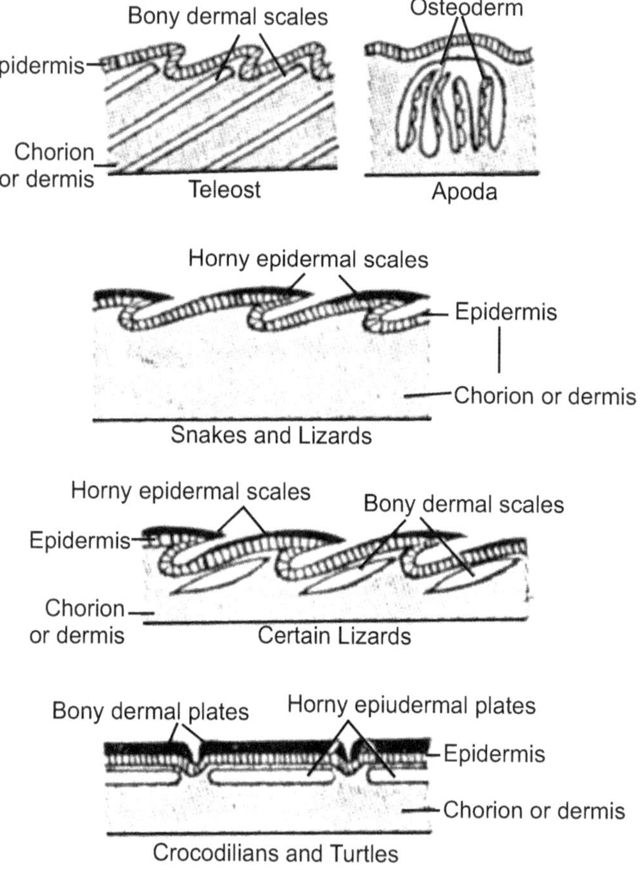

Fig. 1.3: Diagrammatic V. S. through skins of various vertebrates showing relationship of various types of scales

There are small epidermal scales found in birds on the lower leg, foot and base of beak. The thin sheat on the beak called rhamphotheca is also modified stratum corneum.

Some mammals like rat and beavers show scales on feet and tails. There are scales in scaly ant eater and fused plates and bands in *Armadillos* but they do not show *ecdysis*.

1.2.2 Dermal Scales and Scutes

These structures develop in dermis and are mesodermal in origin. In fishes they form exoskeleton. There are five different types of dermal scales found in fishes such as cosmoid, placoid, ganoid, cycloid and etenoid scales.

In some amphibians e.g. Apoda dermal scales or bony plates called osteoderms found embedded in the pockets of dermis. They are also found in back of some tropical toads.

The reptiles also show bony dermal scales. Few lizards, crocodiles, and alligators show bony plates in dermis of their back and neck. In turtles, there are large bony plates forming box like dermal skeleton around trunk including carapace and plastron.

In mammals, bony plates found only in armadillos and whales. Dermal fin rays are present in fishes supporting the fins. They are hair like, bony and formed by fibrous connective tissue.

Digital cornification: All the structures of digits like claws, nails and hoofs are also the modifications of *stratum corneum* at the tips of digits and grow parallel to the skin.

(i) Claws: These structures are found in reptiles, birds and mammals which are identical in structure. Its claw is made up of hard, pointed, curved, narrow, horny plate called *unguis*, and a less hard ventral plate called *subunguis*. Both the structures enclose the tip of the digit covering the last tapering phalanx.

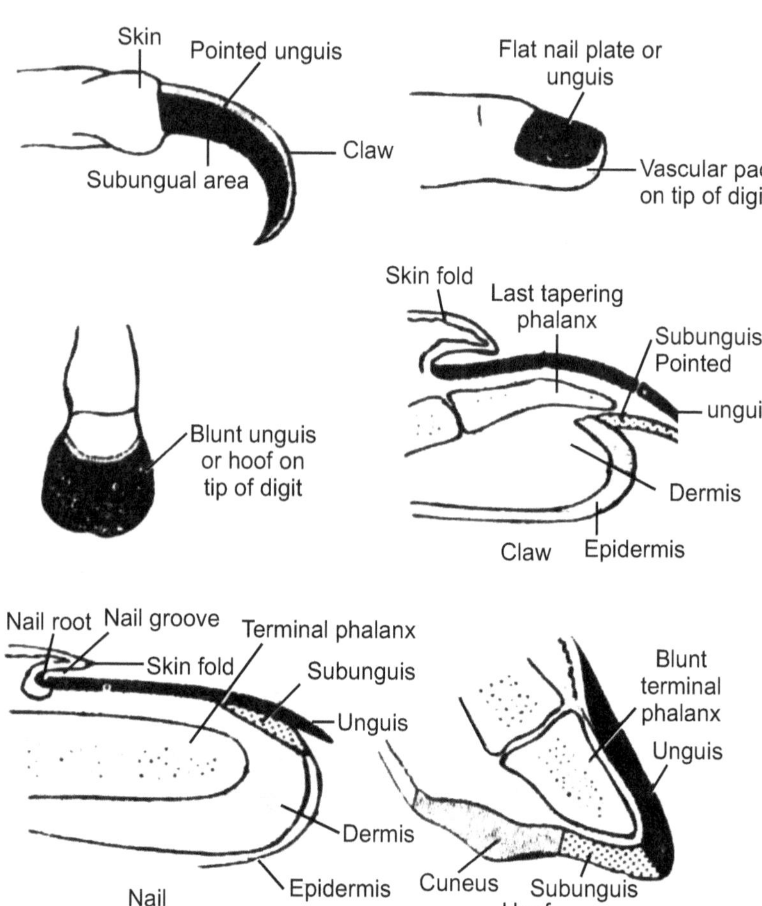

Fig. 1.4: Relation between claw (eagle), nail (human) and hoof (hourse). Digital tips shown complete above and in sagital sections below

(ii) Nails: Actually claws are modified into nails which are characteristic of mammals (primates). In nail, the unguis or dorsal plate is broad and flat, whereas the ventral plate or sub-unguis is much softer and reduced. The tip of the digit forms a greatly sensitive and highly vascular pad. On this pad epidermis invaginates to form a nail groove containing the nail roof.

(iii) Hoof: Hoofs are found only in ungulates. The horny unguis is U or V shaped and very thick. It touches to the ground. Sub-unguis covers the softer horny substance called **cuners**. The tip of the digit form a pad and it contains blunt phalanx. Stratum corneum shows also other modifications such as horns, whale bone plates in toothless whales, horny covering of horns of sheep and cattle and prong horns of antelopes.

Horns: Horns are the characteristic of hoofed mammals only and they are located on head. These are also the organs of offense and defense. Horns are classified into six types but all are not true horns, because they are not formed from stratum corneum. They are true horns, prong horns, antlers, Giraffe horns and hair horns.

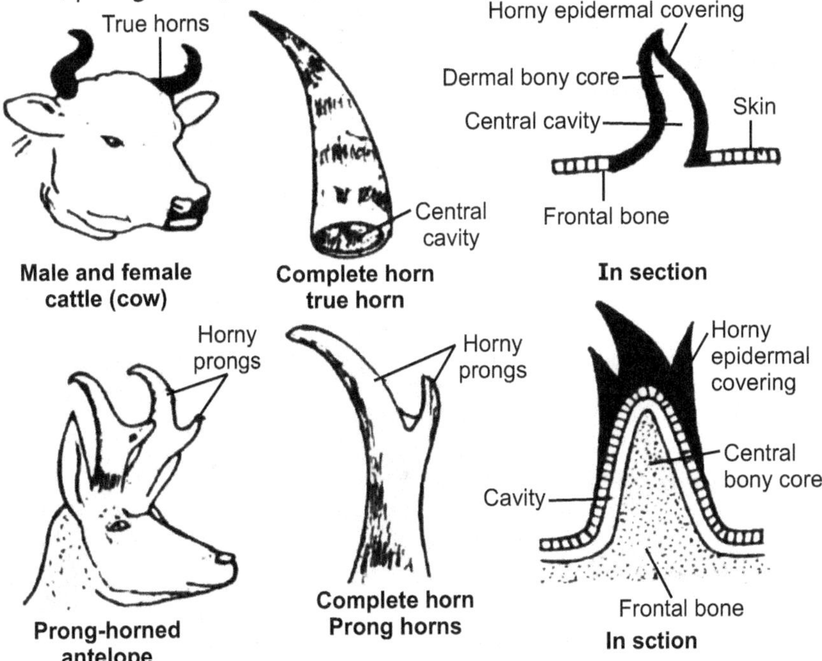

**Fig. 1.5: Integumental and its derivatives,
true horns and prong horns**

(1) True horns: True horns are hollow and they occur in both the sexes of sheep, goat and cattle. True horns are unbranched, cylindrical and tapering. These structures are permanent and show continuous growth throughout life of the animal and are never shed.

The true horn is originated from frontal bone of the skull and it is formed by hollow dermal bony core, and it is covered by an epidermal horny hollow cap.

(2) Prong horns: These are also true horns found in prong horned antelope (**Antilocapra**). These horns are also formed by a small central permanent bony core which arises from frontal bone. This is covered by a thin hollow and horny epidermal horn. The epidermal horny covering or sheath forms 1 to 3 branches or every year. Around the permanent bony core again new epidermal horny prongs are formed during the following year.

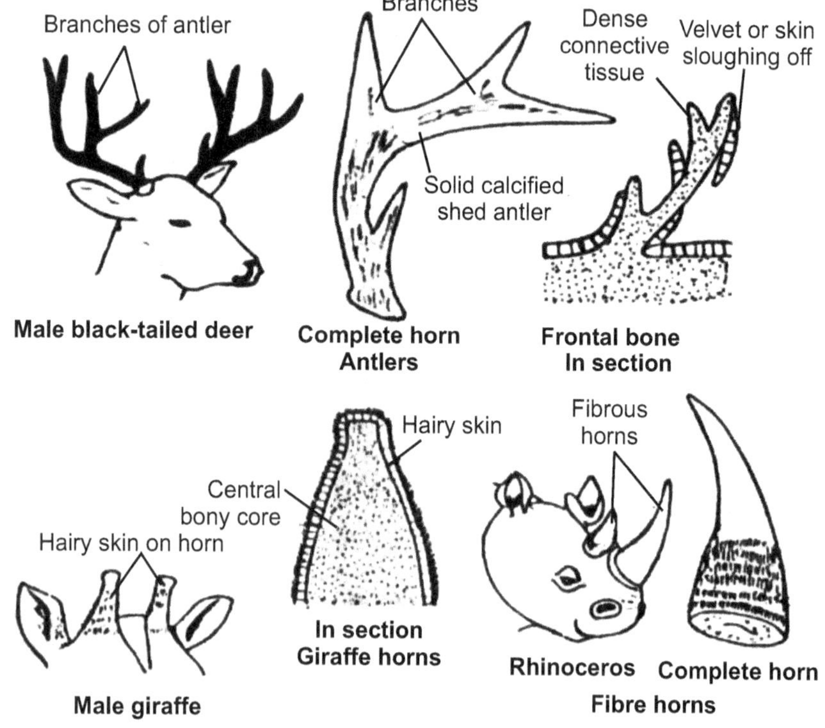

**Fig. 1.6: Integumental and its derivatives,
Antlers, Giraffe horns and Fibre horns**

(3) Antlers: Antlers are found only in deer family. They are seen only in males but formed in both the sexes of reindeer and caribou. Antlers are not true horns but are annual growth. Each antler is a branching solid outgrowth of dense connective tissue connected basally to the frontal bone of skull. The antlers become very hard due to deposition of calcium salts. The antler during its growth covered by a typical hairy skin is fully grown, the velvet covering wears off and antler become naked. After breeding season is over, the antlers are also shed and new antlers develops the following year.

(4) Giraffe horns: These are present in both the sexes. They are short, unbranched and permanent antlers. Each horn has bony dermal core, arising from frontal bone of the skull. This is covered by simple uncalcified skin or velvet which is never shed.

(5) Hair horns: These are also called fibre horns and found in both sexes of rhinoceros. They are located on nasal bones. Their number is variable. Indian rhinoceros has a single horn, while African species has two, one behind the other. These horns are formed by hairy and keratinized epidermal fibres fused together. These horns are permanent structures and again grow if broken. For these horns rhinoceros are still slaughtered illegally because these horns are in great demand in oriented countries as love charm.

Hairs: Presence of hairs on the body is the characteristic of mammals. In some animals, entire body is covered by hairs called *furred* animals. In man, hairs are present in patches and they are scattered in whales. The hair is a solid, slender and elastic horny thread like structure developed from the epidermis. The hairs are placed in deep narrow pits or pockets that traverse the dermis. Each hair consists of a *shaft* that projects above the surface and *roof* embedded in the dermis. At its lower end the root represents a knob like expansion, the hair, bulb, which contains inside a mass of connective tissue called hair papilla. Hair root is surrounded by hair follicle which consists of an epidermal and dermal portion. There are blood vessels and nerves provided to the hair papilla.

The hair is composed entirely of epithelial cells which are arranged in three definite layers, inner medulla, middle cortex and outer cuticle. The cells of the shaft become keratinized, hardened and soon die, so the hair producing above the skin is a dead structure. The hair is kept soft by lubricating the oily secretion of sebaceous glands into hair follicle. A smooth arrestor pili muscles are inserted into the outer most root sheaths of hair follicles. These help in erecting the hairs. The outer most cuticular layer of hair is made up of microscopic scales, the middle cortex consists of several layers of cells. They become slender and keratinized above pigment granules are present in between these cells. Medulla or inner layer contains keratinized cells and intercellular spaces. Hairs are sensory or tactile organs.

Feathers: The feathers are characteristic of birds only because they do not occur in any other group of animals. They are light elastic water proof and most important in flight. There are cornified products of stratum corneum of epidermis. The different colours of the feathers are due to the presence of melanin pigment of various shades as well as due to iridescence. The feathers are present almost throughout the body forming the plumage. The feathers are shed annually, generally late in summer. The new ones develops from the old papillae.

Generally three types of feathers are recognised, countour, down or plumules and hair likes filoplumes.

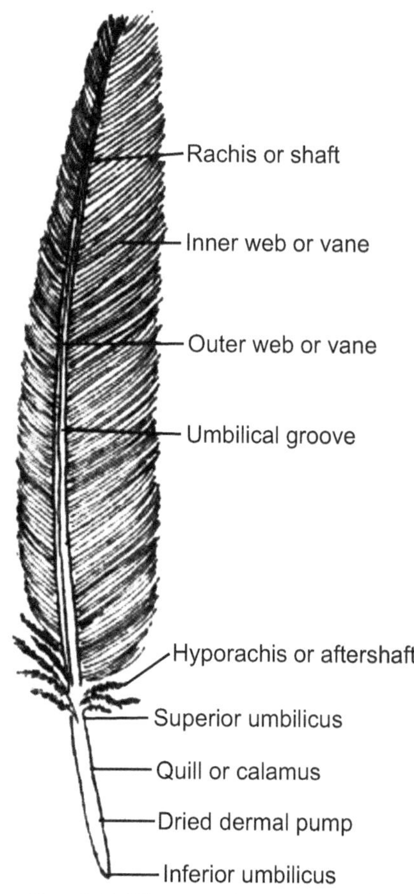

Fig. 1.7: Pigeon. Structure of a typical flight feather (remex) in ventral view

The feathers are developed from the feather papillae or dermal papillae. It grows into a conical cylinder and sinks into depression, the annular groove around the base of the papilla. At the same time, it sinks into a tubular pit, the feather follicle. Papilla has a core of dermal pulp in the centre surrounded by the stratum germinativum and the cornified cells of stratum corneurn.

The cornified layer is known as periderm and the developing feather is called *feather germ*. The actively dividing cells of stratum germinativum become arranged into longitudinal columns that will change into barbs. A feather germ with rudiments of barbs is called a pin-feather and projects well above the skin. Depending on the nature the feather develops into particular type. The feathers perform different functions such as protection, heat retention, protective colouration, organs of flight, se determination and courtship, nest formation, natal covering etc.

Points to Remember

- Scales and scutes are epidermal derivatives.

- Dermal scales and scutes are mesodermal in origin.

- Claws, nails, hoofs, grow parallel in skin.

- True horns and pong horns are types of horns.

- Antlers are found only in deer family.

- Presence of hairs on body is the characteristics of mammals.

- Hairs horns are present on rhinoceros.

- Feathers are present in birds.

- Epidermal glands are derivatives of integument.

- There are ten types of epidermal glands.

Exercise

1. What is integument? Describe the integument and its derivatives in vertebrates.

2. Describe the derivatives of integument in vertebrates.

3. Give an account of epidermal and dermal scales and scutes in vertebrates.

4. Write short notes on:

    (a) Epidermal derivatives

    (b) Dermal derivatives

    (c) Epidermal scales and scutes

    (d) Dermal scales and scutes

    (e) Digital cornification

    (f) Horns

    (g) Feathers

    (h) Epidermal glands.

Chapter **2**...

Skeletal System

Contents ...

2.1 Introduction

The hardened tissues of body together form the skeleton. In vertebrates the skeleton forms a framework of the body to which muscles are found attached. In the body, the skeleton is made up of cartilage or bone or of both cartilage and bone. The presence of endoskeleton is a characteristic of vertebrates, though many vertebrates also have dermal exoskeleton.

2.2 Types of Vertebrate Skeletons

There are three types of skeletons which develop in vertebrates.

(1) **Epidermal horny exoskeleton:** These include hard and horny or keratinized derivatives of epidermal layer of skin, such as claws, reptilian scales, bird feathers and mammalian hairs, horns, nails and hoofs etc. There is no exoskeleton in amphibians.

(2) **Dermal bony skeleton:** It is derived from dermis of skin. It includes bony scales, plates, or scutes, fin rays and antlers of fishes, reptiles and mammals.

(3) **Endoskeleton:** It develops from mesenchyme. At early embryonic stage, endoskeleton is composed of cartilage, which is replaced by bone in most adult vertebrates.

2.3 Functions of Endoskeleton

(1) It forms a framework for the support of the body.

(2) It permits growth because it is living and grows in size with the rest of the body.

(3) It maintains a definite shape and form of an animal.
(4) It provides protection for delicate vital organs of the body.
(5) It provides a firm and adequate surface for attachment of muscles by means of tendons.
(6) In bone marrow, blood corpuscles are manufactured.
(7) Ear ossicles aid in hearing.
(8) Tracheal rings and ribs help in breathing.

The endoskeleton of vertebrates is divided into three types as follows:

(1) Somatic skeleton: Skeleton of body wall.
(a) Axial skeleton: Vertebral column, ribs, sternum and skull.
(b) Appendicular skeleton: Limbs and girdles.
(2) Visceral skeleton: Skeleton of pharyngeal wall (splanchnocranium).

2.4 Appendicular Skeleton in Vertebrates

Appendicular skeleton includes pectoral and pelvic girds and appendages. Appendages are median and paired. Median appendages are found in the fishes and aquatic tetrapoda and paired appendages are found in all vertebrates except cyclostomes. In case of fishes, the paired appendages are pectoral and pelvic fins, in tetrapoda they are fore and hind limbs.

Girdles and Limbs in Tetrapoda:

Girdles and paired limbs of tetrapoda are similar and have homologous parts.

Pectoral girdle is made up of several cartilage bones. They are a scapula, a suprascapula, pre-coracoid and a coracoid in each half. There is a glenoid cavity between scapula and coracoid. Two dermal bones, the clavicle and interclavical are added in each half. But there is a gradual reduction or loss of some bones.

Pelvic girdle has three cartilage bones in each half. They are ilium and pubis. These bones form an acetabulum cavity. In mammals acetabulum is formed by cotyloid or acetabular bone. Generally, a large obturator formaen lies in between the ischium and pubis. The two halves of pelvic girdle often meet midventrally at the pubic or ischiac symphysis. In most tetrapoda, the pelvic girdle is attached to the axial skeleton through the ilia articulating with sacral vertebrae.

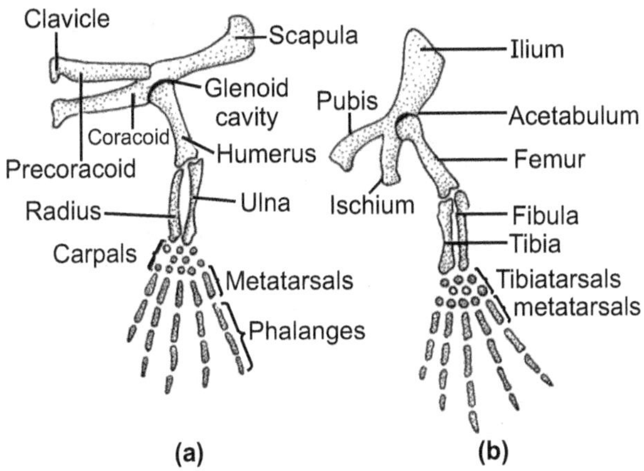

Fig. 2.1: Girdles and Limb Bones of Tetrapoda

(a) Pectoral Girdle and Forelimbs, (b) Pelvic Girdle and Hindlimbs

Limbs: The limbs of tetrapoda are pentadactyl. They are used for locomotion and to support the body weight above the ground. The limbs are with joints and three segments, the stylopodium, zeugopodium and autopodium are present with joints between the segments. The stylopodium is upper arm (or thigh), having a single humerus (or femur). The humerus joins the pectoral girdle at the glenoid cavity, while the femur joins the pelvic girdle at the acetabulum cavity. The zeugopodium is the forearm (or shank) which contains two parallel bones, the radius and ulna (or tibia and fibula). The radius and tibia are lateral, while ulna and fibula are medial. Between the stylopodium and zeugopodium is an elbow joint (or knee joint). The autopodium has three divisions, a carpus or wrist (tarsus or ankle), metacarpus or palm (metatarsus or sole), and digits forming fingers or toes. The carpus or tarsus in the primitive condition has 10 bones in three rows, the first row has a radiale (or tibilae), an intermedium, and ulnare or fibulare), in the second row are generally two Centralia, while the third row has five distal carpals (or distal tarsals). The metacarpus (or metatarsus) has five long metacarpals (or metatarsals). The digits are generally five in number or pentadactyl, with the first digit on preaxial side and the fifth on the post axial side. Each digit has a linear row of phalanges, they are 2, 3, 4, 5, 4 beginning from the first to the last. There are some evidence that are early tetrapoda had seven digits, one on the pre-axial side was prehallex (or prepollex), and the post axial digit was a post minimus. But these were additional bones of the wrist or ankle.

Girdles and Limbs of Tetrapoda - Comparative Account:

1.   Amphibia (Frog): Pectoral girdle is an arch enclosing the chest and two halves joined mid-ventrally and closely related to the sternum. Each half has coracoid, precoracoid, clavicle, paraglenoid cartilage, scapula and supra scapula. Glenoid cavity is bordered by coracoid paraglenoid and scapula. Anura have two kinds of pectoral girdles, firmisternal and arciferal. Sternum is fused to the middle of pectoral girdle. The pectoral girdle is reinforced anteriorly by an omosternum, posteriorly by mesosternum and their cartilaginous expansions, the episternum and xipisternum respectively.

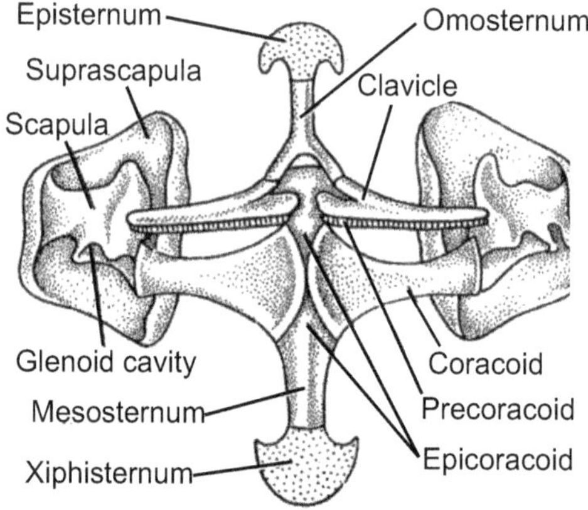

(a) Pectoral Girdle and Sternum (Ventral View)

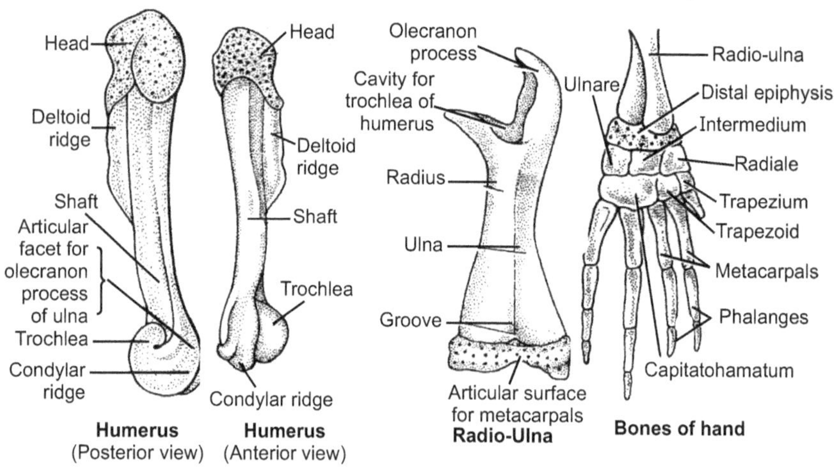

(b) Forelimb Bones

Fig. 2.2: Pectoral Girdle of Frog and Forelimb

Forelimb: Humerous has a prominent deltoid ridge. Redius and ulna fused into a radiulna. These are five carpals into two rows. In the first row the intermedium and ulnare are fused together and there is a radiale. There is a single centrale. The three carpals of the second row are fused together. Thus, member of carpals is reduced. Only four digits are present besides a rudimentary pollex enclosed in the skin.

Pelvic Girdle: It is V-shaped and has a long ilium on each side. It is joined with the transverse processes of the sacral (9th) vertebra. The ilia act as a shock absorbers when the animal lands after a jump. Pubis and ischium are small. A solid fulcrum is formed by the fusion of two pubes and two ischia. All three bones join at the acetabulum cavity.

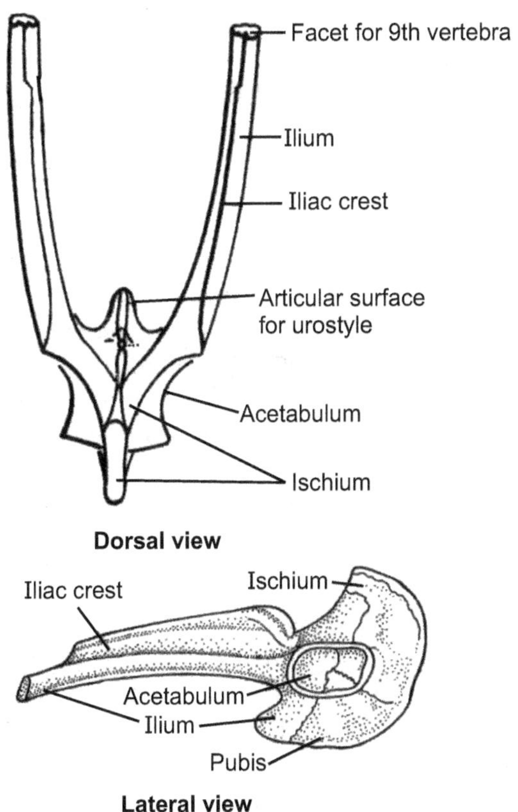

(a) Pelvic Girdle

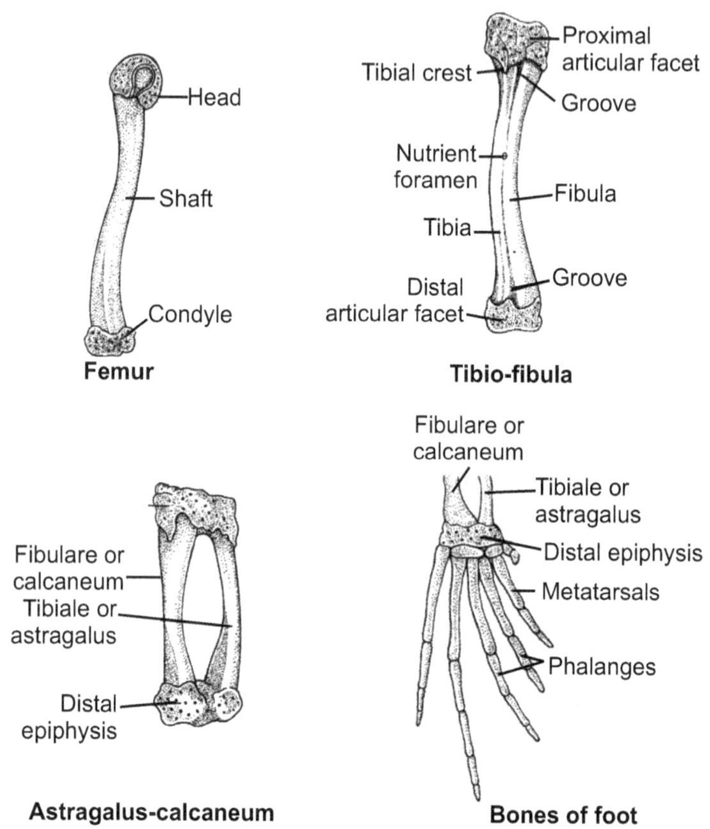

(b) Hind Limb Bones
Fig. 2.3: Pelvic Girdle of Frog and Hind Limb

The pelvic girdle of frog is different from other vertebrates. It has long ilia forming long lever for transferring the force from hind limbs to the vertebral column in jumping.

Hind Limb: The tibia fibula fuse into tibiafibula. There are five tarsals in two rows, the proximal row has two long bones an inner astragalus and outer calcaneum which make ankle very long for jumping. There are five digits and first one is called *hallux*. There is an additional digit called calcar or prehallux enclosed in the skin. It helps in stretching the web of the foot.

2.   Reptilia (Varanus): In reptiles, the girdles and limbs are well developed except in snakes and limbless lizards, but vestigial pelvis and hind limbs are found in some snakes.

Pectoral Girdle: It is well developed and joined to sternum. Each half has a bony coracoid and scapula, between which is a glenoid cavity. Supra scapula is calcified cartilage. Epicorcoid cartilage in between coracoids. In each half, there are three fenestrae between coracoid and epi-coracoid. T-shaped unpaired dermal bone called interclavicle (episternum) lies midventrally. Two clavicle bones are present. Attached to the coracoids is a sternum made of cartilage.

Forelimb: It is typical with no unusual feature. Humerus is broad, radius and ulna are separate. Carpals are in two rows : four in the first row including a centrale and five in the second row. There are five digits ending in claws.

Pelvic Girdle: All three bones are fused in separably to form a single innominate bone in each half. Both pubis and ischium meet their fellows to form pubic and ischiac symphyses. The pubis has a small foramen for the obturator nerve. Between two innominate bones is a large ischiopubic foramen divided by a ligament into two obturator foramina. Acetabulum is bordered by all three bones.

Hindlimb: It is typical, but the tibia begins to become larger than the fibula. Tarsals are reduced by fusion. The proximal row has two bones and distal row has three between two rows of tarsals is an intratarsal ankle joint where bending takes place. There are five digits ending in claws.

3. Aves (Gallus): In case of birds, pectoral girdle and forelimbs are modified for flight. The pelvic girdle and hindlimbs are modified to support the weight of body in standing and walking. Bones are spongy, tubular and pneumatic so that they are light.

Pectoral Girdle: It has a large, stout coracoid attached to the sternum at the coracoid groove by means of immovable articulation. Coracoid is joined at its other end at right angle with a sword like scapula laying over and bracing the ribs. Glenoid cavity is formed between scapula and coracoid. Thin clavicle is attached to each coracoid.

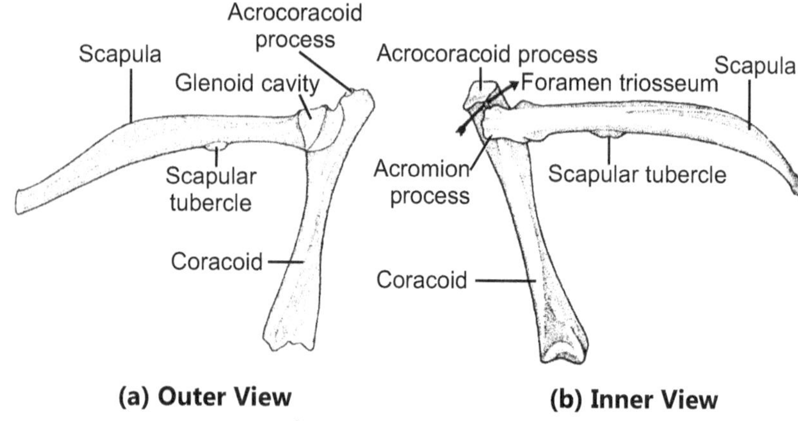

(a) Outer View   **(b) Inner View**

Fig. 2.4: Pectoral Girdle

Two clavicles are joined ventrally to a small, flat interclavicle forming a furcula. In flying birds, the furcula is joined to coracoids and is well developed. In non-flying birds, it is reduced or absent.

Forelimb: Its skeleton is modified for flight. Humerus is short with deltoid ridge. Redius and ulna are well formed. Carpus has four carpals in two rows - in the first row are radiale and ulnare, but the carpals of the second row fuse to three fused metacarpals to form a carpometacarpus. First metacarpal is much reduced, all three are fused proximally, but distally the second and third are free forming two rods. There are three digits, first and third has one phalanx and second has two phalanges. The digits are reduced to almost a one fingered condition. Muscles of upper arm are large and strong, those of forearm are reduced and those of the hand have atrophied.

Pelvic Girdle: Ilia and ischia have become long and wide and lie longitudinally. Ilium extends infront and behind the acetabulum and fused to the synsacrum along its entire length. Ischium is broad and extends posteriorly. A slender pubis extends backwards and ends freely. Between the ischium and pubis is an obturator foramen.

Hindlimb: These are modified for bipedal locomotion. A strong femur articulates at the acetabulum. Fibula is much reduced and is short. A sesamoid bone forms patella. The tibia is fused with the proximal tarsals to form tibiotarsus. The distal tarsals fuse with the second, third and fourth fused metatarsals to form compound bone, the tarsometatarus.

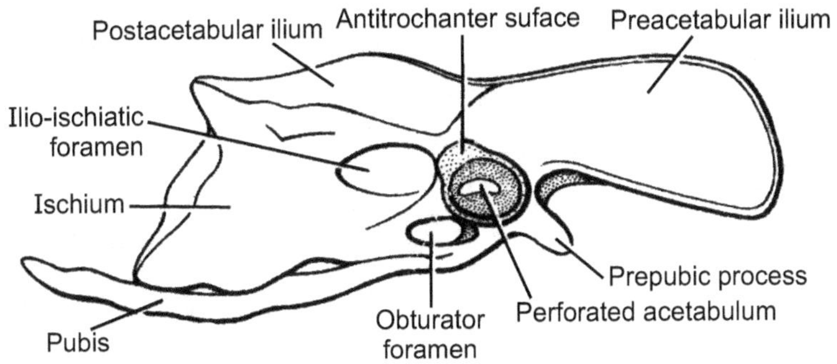

Fig. 2.5: Pelvic Girdle and Hindlimb of Gallus

There is a bony spur projects near to the lower end of the tarsometatarsus. It is useful in male birds for fighting. The digits are never more than four, having 2, 3, 4, 5 phalanges. The last phalanx of each digit bears a claw. The first digit (hallux) usually points backward, while the other three digits are directed forward.

Mammalia (Rabit): Each half of the girdle has triangular scapula or shoulder blade, at the end of which is glenoid cavity. Near this cavity there is corcacoid process formed by the reduced coracoid. On the outer surface of the scapula is a spine which has two projections acromion and metacromion process for attachment of muscles. There is clavicle. They are present in mammals which have an extensive movement of forelimbs, in others they are reduced or lost.

Forelimb: Humerus is stout with deltoid ridge. The distal end has two expanded condyles between which is a pulley or trochlea for articulation with ulna. Radio and ulna are separate but tightly bound together. The ulna is produced into an elbow or olecranon process, in front of which is a sigmoid notch for articulation with the humerus. Carpus has eight bones, three in first row, then a small centrale and four distal carpals. There are five metacarpals having five digits with 2, 3, 3, 3, 3, phalanges ending in claws. In primates the first digit or pollex is independent of the others. It can be brought opposite the other digits and palm.

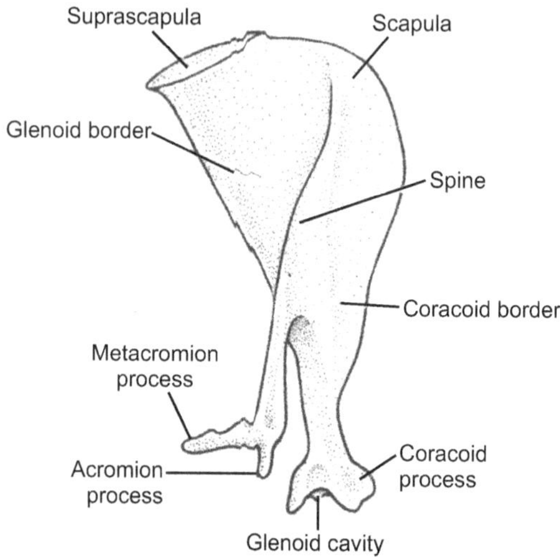

Fig. 2.6: Rabbit. Pectoral Girdle. Right Half in Outer View

Pelvic Girdle: In each half the ilium, ischium and pubis fuse to form an innoninate bone. The two halves meet by a midventral pubic symphysis. In rabbit and some mammals there is small acetabular (cotyloid) bone. Acetabulum lies where all three bones meet. On ecach side, there is a large obturator foramen between pubsis and ischium. The two ilia join with the transverse processes of the first vertebra of the sacrum.

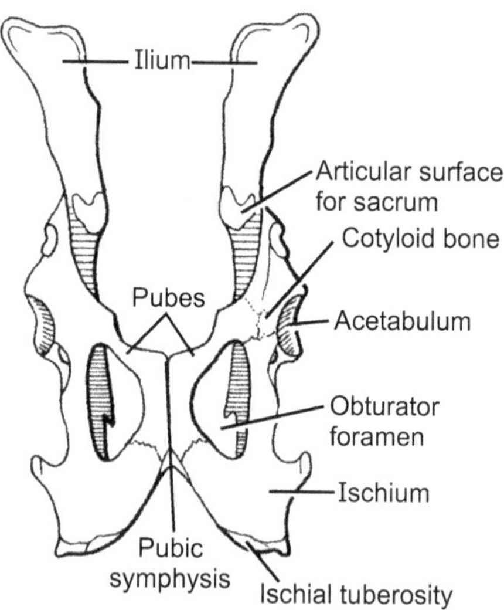

Fig. 2.7: Rabbit. Pelvic Girdle

Hind limb: Femur is strong with the head for articulation with the acetabulum. Tibia is strong bone and patella forms knee cap. Fibula is reduced and fused distially to the tibia but separate proximally. Tarsus has six bones, proximal row has astragalus and calcanium. There are four metatarsals bearing digits. Each digit has three phalanges ending in claws. First digit or hallux is absent. In terrestrial mammals, there are three kinds of postures of hands and feet like plantigrade (entire hand and foot in contact with ground e.g. bear, rabbit), digitigrad (only digits are placed on ground e.g. dog, cat) and unguligrade (only tips of digits are placed on ground e.g. deer, horses, cattle).

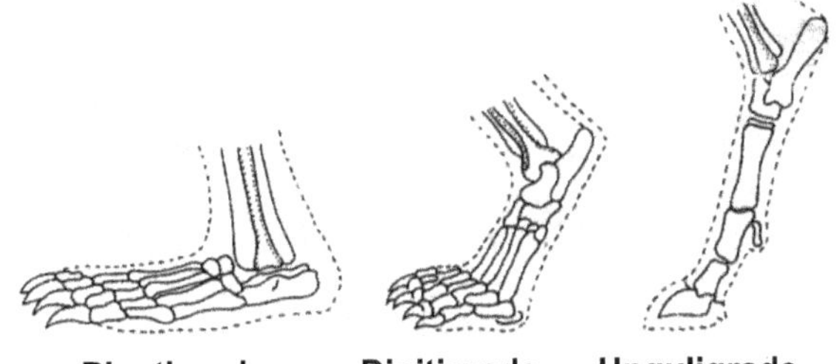

Fig. 2.8: Foot Postures in Locmotion of Mammals

2.5 Axial Skeleton in Vertebrates

Axial skeleton is formed by skull and vertebral column in vertebrates.

Skull: The skeletal framework of the vertebrate head is called a skull. It protects the delicate organ like brain. Skull is derived from three different embryonic components like chondrocranium, splanchnocranium and dermatocranium.

1. **Cyclostomata:** Skull in cyclostomes is cartilaginous and very incomplete. In hag fishes, it is made up of cartilage plate on which brain rests. Optic capsules are joined to posterior part of this plate. In case of lamprays roof is absent and brain is protected by fibrous connective tissues. Jawa are lacking in cyclostomes.

2. **Chondrichthyes (Cartilaginous Fishes):** In these fishes, skull is cartilaginous and brain is roofed. Olfactory and otic capsules are

present in skull. No bone present. The first mandibular ach forms upper and lower jaws. Upper jaws is fused with skull.

3. Osteichthyes (Bony Fishes): Dermal bones cover the roof of the skull. In higher bony fishes many dermal bones are present and they form armour around the skull. Parts of chondrocranium are replaced by bone in occipital, sphenoidal, otic and ethmoid regions. Jaw suspension in bony fishes may be hyostylic or autostylic.

4. Amphibia: There is reduction of bones in the skull of amphibians and skull becomes flat. Many bones like premaxillaries, maxillaries, nasals and frontals, parietals and squamosals are paired. The dorsal surface of the skull is solid and covered by dermal bones with openings for nares and eyes. Exoccipitals are ossified and bear two occipital condyles on each exoccipital. Epipterygoid is never ossified in amphibians and incompletely ossified in urodeles and fused with otic region. Articular lower jaw is bony in candates and cartilaginous in anurans.

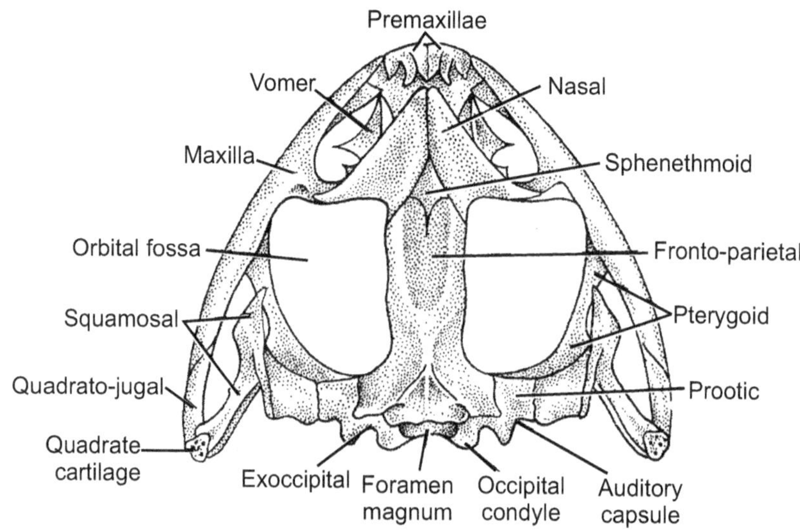

Fig. 2.9: Skull of Frog

5. Reptilia: There are more dermal bones than amphibias. The parietals are fused together and interparietal foramen is prominent in lizards. The temporal fossae are present for muscles of the jaws. The occipital condyle is single. A stout ectopterygoid is added between maxilla and pterygoid. In lower jaw in each half has one cartilage bone and five dermal bones. Meckel's cartilage becomes very slender. The quadrate bone forms a movable union with squamosal. Jaw suspension is streptostylic.

6. Aves: The skull of bird is larger and very light due to pneumatic bones. Bones are fused together. There is a large, arched cranium due to more development of the brain. The orbits are immense and lie in front of the cranium. They are separated by thin interorbital septum. Orbits are very large and continuous with temporal fossa. Each orbit is surrounded by dermal bones called sclerotics. There is a large pointed beak formed by the premaxillae and dentaries. Teeth are absent. Occipital condyle is single. Foramen magnum is surrounded by four occipital bones.

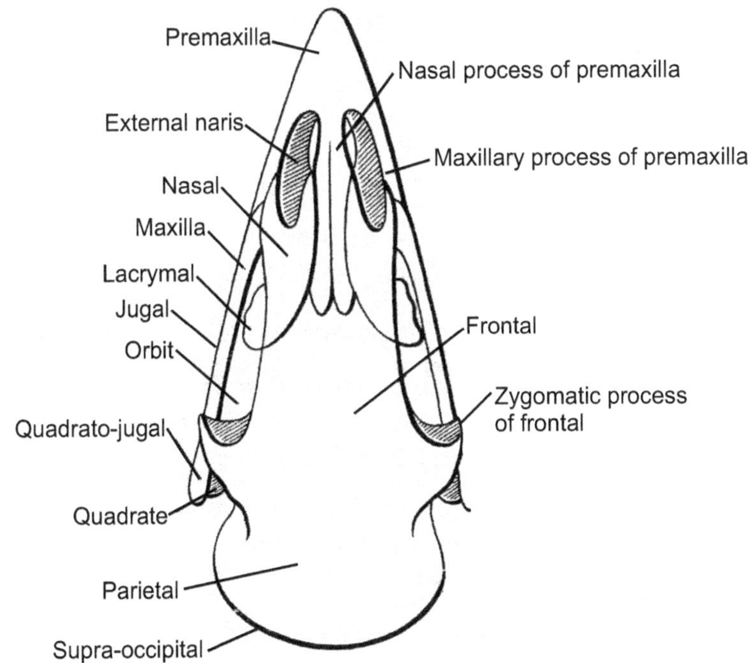

Fig. 2.10: Skull of Aves

Meckel's cartilage is much reduced. The jaw suspension is streptostylic in which lower jaw has to movable joints. Hyomandibular forms the columella of the middle ear.

7. Mammalia: There is much variation formed in skulls of different mammals. Skull bones are reduced by loss or fusion. The prefrontals, postorbitals and quadratojugals are lacking. In mammals four occipital bones are fused together. There are two occipital condyles. There is fusion in sphenoidal area. Many bones form a single bone. Due to long brain, the cranium is very large expanded dorsally and laterally. Ear ossicles, incus, malleus and stapes are formed. Lower jaw is in each half is made of a single dentary, with no traces of Meckel's cartilage. The hyoid arch forms a hyoid apparatus.

Vertebral Column: The vertebral column is made of a metameric series of vertebrae extending longitudinally from the skull to the top of the tail.

1. **Vertebral Column in Fishes:** There are two types of vertebrae namely trunk vertebrae and caudal vertebrae found in fishes. The vertebrae are amphicoelous in which the centrum is concave at both ends. It is most primitive type found in fishes. There are different types of tail fins depending on the shape of the tail. They are protocercal, heterocercal, diphycercal, homocercal and hypocercal.

2. **Vertebral Column in Tetrapoda:** There were many changes appeared in the tetrapoda as they emerge from water to land. "The vertebral column was differentiated into different region like anterior trunk vertebrae called cervicals. Last few trunk vertebrae called sacrals which articulate with pelvic girdle. The caudal vertebrae are reduced in number. Thus, tetrapoda have cervical, trunk, sacral and caudal vertebrae. In many reptiles, birds and mammals the trunk vertebrae are differentiated into thoracic vertebrae with ribs and lumber vertebrae with reduced or no ribs.

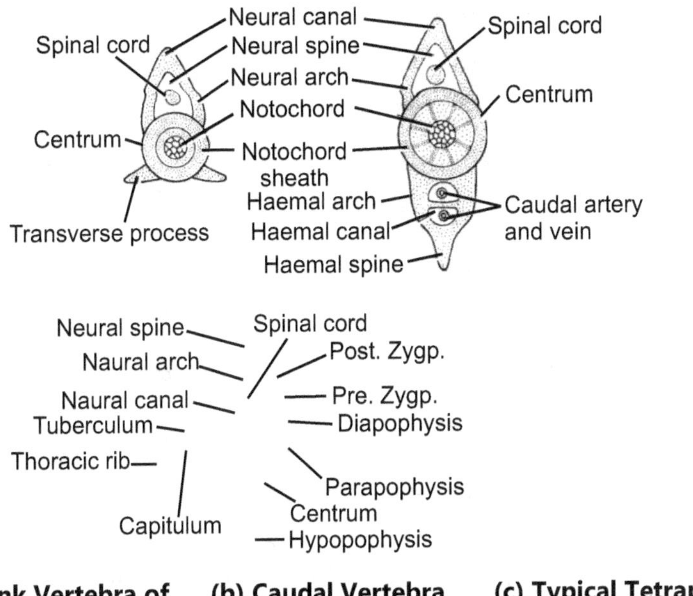

(a) Trunk Vertebra of Shark   **(b) Caudal Vertebra of Shark**   **(c) Typical Tetrapod Vertebra**

Fig. 2.11: Structure of a Vertebra showing processes in Cephalic View

1. **Amphibia:** The centra of vertebrae are stereo-spondylous. There is a single cervical vertebrae (atlas) which articulates with the skull. This is followed by trunk vertebrae, then there is a single sacral vertebrae connected to pelvic girdle. Caudal vertebrae are found only in tailed forms. All caudal vertebrae except the first, bear haemal arches with haemal spines.

2. **Reptilia:** The vertebrae are gastrocentrous. The vertebral column in lizards and crocodilians is differentiated into cervical, thoracic, lumber, sacral and caudal regions. In all living forms, two sacral vertebrae fuse to form a sacrum whose transverse processes and ribs are attached to the pelvic girdle. Caudal vertebrae possess a small V-shaped chevron bones on ventral side of centrum.

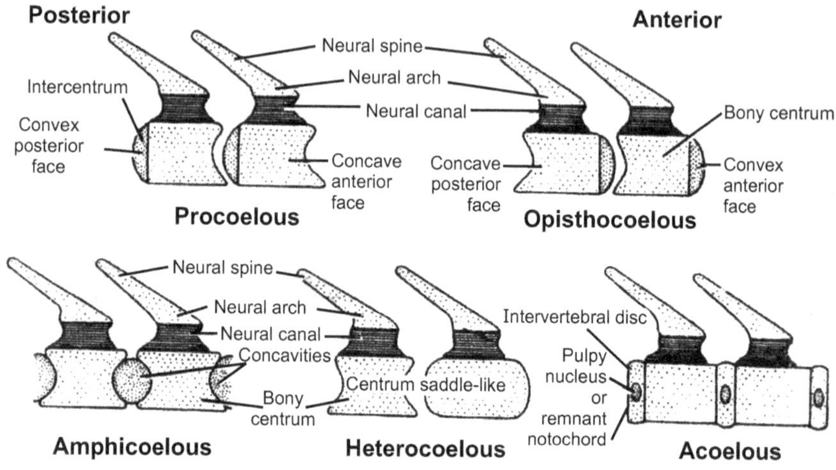

Fig. 2.12: Types of Vertebrae Based on Shape of Centra in Sagittal Section

All types of centra are found in reptiles. They are amphicoelous and some have opisthocoelous vertebrae. Most lizards, snakes and crocodiles have procoelous vertebrae. Sacral vertebrae of crocodilians are amphiplatyan. In snakes and some lizards in addition to zygapophysis there is special pair of apophyses called zygospherus at the anterior end fitting into concavities called zygantra located on the posterior side of each vertebra.

3. Aves: The vertebral column is rigid due to an ankylosis between vertebrae. It is useful in flight. Cervical region is very mobile but other regions are capable of little movement. The vertebral column provides fulcrum for wing strokes. In the cervical, the number of vertebrae is variable from 8 to 25. These vertebrae are heterocoelous or saddle shaped. The anterior face of the centrum is convex dorsoventrally but concave side to side. Cervical vertebrae bear short ribs, ventrally each vertebra has hypapophysis. Thoracic vertebrae are more or less fused together, though some are free, their centra are produced below into hypapophyses and their broad transverse processes bear ribs. Ribs are attached to thoracic vertebrae and are united ventrally to the sternum. The posterior thoracic, lumber, sacral and first few caudal vertebrae are all fused together to form a single synsacrum, which becomes fused with the pelvic girdle. Synsacrum is formed from 13 to 20 vertebrae in different birds. The synsacrum provides a rigid framework valuable in flight and distribution of body weight on the legs in standing or walking. The number of caudal vertebrae is small. First few after synsacrum are free but the posterior ones are fused to form a pygostyle which supports the tail. The free caudal vertebrae are amphicoelous.

4. Mammalia: The vertebrae are gastrocentrous, between the centra are intervertebral discs. The vertebral column has cervical thoracic, lumber, sacral and caudal regions. Important feature is the presence of seven cervical vertebrae in most mammals except in some edentata which have six to nine vertebrae. In many whales, moles and armadillos the cervical vertebrae fuse together. Ribs are fused to cervical vertebrae. Thoracic vertebrae bear ribs which connect directly or indirectly with the sternum, which do not reach the sternum called floating ribs.

Lumbar vertebrae are strong and longer than others. They have prominent transverse processes directed forward. Sacrum has several vertebrae fused together, 3 in dog, 4 in rabbit, 5 in horse and 5 in man. There is no sacrum in whales because of reduced pelvic girdle. Caudal vertebrae are variable in number : 16-18 in rabbit, 20-23 in dog, 15-21 in horse and 3-4 in man which unite to form a single coccyx. Usually, at the tip of tail vertebrae are reduced to only centra, but at the anterior vertebrae bear all the parts of vertebrae.

Sternum: It is an elongated structure in midventral region of the thorax. It is absent in fishes and seen in tetrapoda. In tetrapoda the sternum articulates with the pectoral girdle. In reptiles, birds and mammals, it is articulated with ribs to form a thoracic basket which protects delicate organs and helps in respiration.

1. **Fishes:** Sternum is absent in fishes.

2. **Amphibia:** In amphibian is shows bony omosternum, cartilaginous episternum and bony mesosternum and expanded cartilaginous xiphisternum.

3. **Reptilia:** Sternum is absent in turtles, snakes and most limbless lizards. In others, sternum is rhomboidal plate between coracoids. Three pairs of thoracic ribs are attached to it. T-shaped dermal bone, the episternum lies ventral to the sternum, but it is not the part of sternum.

4. **Aves:** In birds, it is triangular bone extending far posteriorly below the abdomen. Carina or keel is the mid ventral portion. Pair of coastal processes which project anteriorly attached to thoracic ribs. There are two pairs of xiphoid processes. Running birds have a rounded sternum with no keel.

5. **Mammalia:** It has three regions namely presternum, middle mesosternum consisting of five pieces called sternebrae, the fifth is very small and posterior region is called metasternum and it has expanded xiphoid cartilage. Seven pairs of ribs are joined to the sternum in such a way that each rib articulates with two sternebrae. Bats have carinate sternum.

Points to Remember

- Skeletal system supports the body of animal.

- Skeletal system is divided into two types, namely appendicular skeleton system and axial skeleton system.

- Appendicular skeleton includes pectoral, pelvic girdles and appendages.

- Appendages are paired and they are fore and hindlimbs.

- Pectoral girdle is made up of cartilaginous and bony materials.

- In forelimbs humerus, radio ulna and carpels, meta tarsals and phalanges are present.

- In hind limbs femur, fibula, tibia, tarsals, metatarsal and phalanges are present.

- Glenoid cavity is present in pectoral girdle for attachment of forelimbs.

- A cetabulum cavity is present in pelvic girdle for attachment of hind limbs.

- Axial skeleton is formed by skull and vertebral column in vertebrates.

- In cyclostomes and cartilaginous fishes, skull is formed by cartilage and in bony fishes formed by bony material.

- Skull protects brain, eyes and ears.

- Vertebral column is formed by different types of vertebrae.

- The particular bones form the skull.

- Sternum is articulated with pectoral girdle.

Exercise

1. Give an account of appendicular skeleton in vertebrates.

2. What is pectoral girdle ? Describe the pectoral girdles found in vertebraes.

3. What is pelvic girdle ? Give an account of pelvic girdles found in vertebraes.

4. Give an account of structure of forelimbs found in various vertebrae animals.

5. Give an account of structure of hindlimbs found in different vertebrae groups.

6. Write short notes on :

    (i)   Pectoral girdle of frog and lizard.

    (ii) Pelvic girdle of frog and bird.

    (iii) Forelimbs of birds.

    (iv) Forelimbs of reptiles and mammals.

    (v)   Hind limbs of birds and mammals.

    (vi) Hind limbs of reptiles and birds.

7. Give an account of skull in different vertebraes.

8. What is vertebral column ? Write an essay of the vertebral column of the vertebrae groups.

9. Write short notes on :

    (i)   Vertebral column of birds and mammals.

    (ii) Vertebral column of reptiles and birds.

    (iii) Vertebral column of amphibians.

    (iv) Vertebral column of mammals.

Chapter 3...

Digestive System

Contents ...

3.1 Embryonic Digestive Tract

Archenteron: The embryonic archenteron becomes the living of the adult digestive tract and of all its derivatives. Splanchnic mesoderm adds layers of connective tissue and smooth muscles around the archenteron. Ectodermal invagination of the head forms the *stomodaecum* leading into oral cavity and a similar midventral ectodermal invagination forms *proctodaeum*, which leads into the hindgut. The adult buccal cavity is formed by stomodaeum and gives rise to teeth enamel, epithelial covering of tongue, glands. e.g. mucous, poison and salivary etc. and Rathke's pouch of anterior pituitary gland. The proctodaeum forms either a small terminal part of the cloaca in lower vertebrates and rectum in mammals.

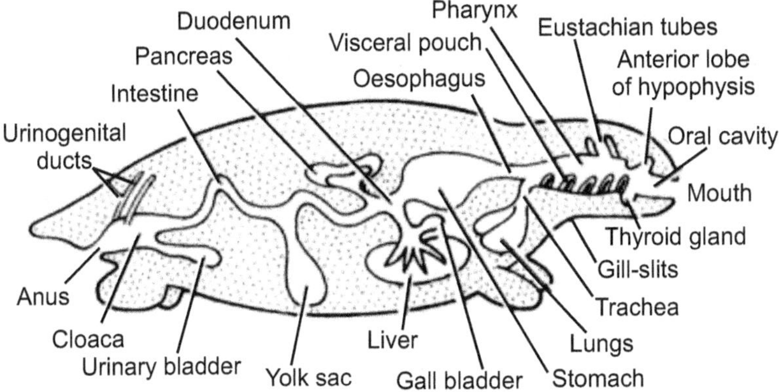

Fig. 3.1: Alimentary canal and its chief derivatives in a vertebrate

The alimentary canal in embryos from stomach to cloaca is attached to the dorsal body wall by a double fold of peritoneum, called *dorsal mesentery* and to ventral body wall by *ventral mesentery*. In adults, dorsal mesentery persists but the ventral mesentery disappears leaving only in the region of liver and urinary bladder.

3.1.1 Digestive Tract of Adult

The digestive tract differentiates for different functions in the following regions: mouth, buccal cavity, pharynx, oesophagus, stomach, small intestine, large intestine and cloaca. Following outgrowths arise from the digestive tract: Oral glands, Rathke's pouch, thyroid gland, gill clefts, tympanic cavity, thymus and other glands of gill clefts, trachea, lungs, swim bladder, liver, pancreas, yolk sac and urinary bladder.

(1) Mouth: Mouth is the anterior opening leading into oral cavity. In lamprays (cyclostomes) it is circular opening at the base of buccal funnel and remains permanently open due to lack of jaws. In gnathostomes it is terminal. Mouth is bounded by lips which are immovable and formed of cornified skin in fishes, amphibians and reptiles. In mammals, these are fleshy and muscular.

(2) Buccal cavity: The space between the lips and the jaws is a vestibule. It may be bounded on the outside by cheeks and on the inside by the gums. Mucous glands of cheeks open into the vestibule. The mouth opens into the buccal cavity, which is the space between the mouth and the pharynx.

In elasmobranchs and most bony fishes, the nasal cavities do not open into the buccal cavity. In amphibians and tetrapoda, the nasal cavities open into the buccal cavity by *internal nares*, which are primitively placed anteriorly, but in crocodiles, birds and mammals, they become posterior in the pharynx due to the formation of a *secondarly palate* and it separates respiratory nasal passage from the mouth cavity or food passage. In birds, this palate is cleft due to which nasal and buccal cavities communicate with each other. In mammals, secondary palate is continued posteriorly as a membranous soft palate. In human beings *soft palate* hangs into the laryngeal pharynx in the form of fleshy process, called *uvula*.

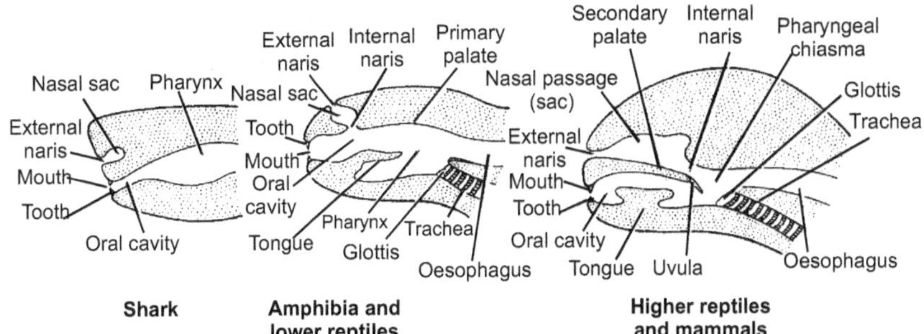

Fig. 3.2: Relation of nasal passages to buccal cavities in different vertebrates

Derivatives of Buccal Cavity: These are mainly teeth, tongue, oral glands and anterior and middle lobes of pituitary (adenohypophysis).

(1) Teeth: These are hard and pointed structures attached to jaw bones that aid in food-getting. In vertebrates, two types of teeth occur called epidermal and true teeth. Epidermal teeth are horney projections of stratum corneum and best represented in cyclostomes. Other examples are conical projections from lips of tadpoles of some species of frogs, serrations on beaks of some turtles and birds, horny plates in ducks bill, sirenians and baleen whales and egg tooth for cracking eggshell before hatching in turtles, *sphenodon*, crocodiles, birds and monotremes.

True teeth occur in all vertebrates except agnathans, sturgeons, some toads, siren, turtles modern birds etc. Teeth are poly phyodont, acrodont and homodont in fish, amphibians and most reptiles, but they are diphyodont, thecodont and heterodont in mammals. Teeth are similar in structure to the placoid scales of sharks, composed of a core of dentine surrounded by a crown of enamel and are believed to have evolved from bony scales.

(2) Tongue: The tongue is found mostly in all vertebrates. There is great diversity seen in tongue. In cyclostomes, the tongue is muscular, fleshy and rasping with horny teeth. Fishes show primary tongue formed of a fleshy fold of the buccal floor. There are no muscles in the tongue but receptors and teeth are present on the tongue in some bony fishes. The tongue is covered with mucous membrane. In some amphibians tongue is absent or immovable. Most amphibians show protrusible tongue and folded back on itself when not in use. It can be thrown out of mouth by rapid inflow of lymph for capturing the insects. The tongue in lizards and snakes is often highly developed. In chameleons it is very extensible used to capture insects. The tip is thickened and sticky. Snakes bear forked tongue and useful for external environment stimuli. In turtles and crocodiles, the tongue cannot be extended.

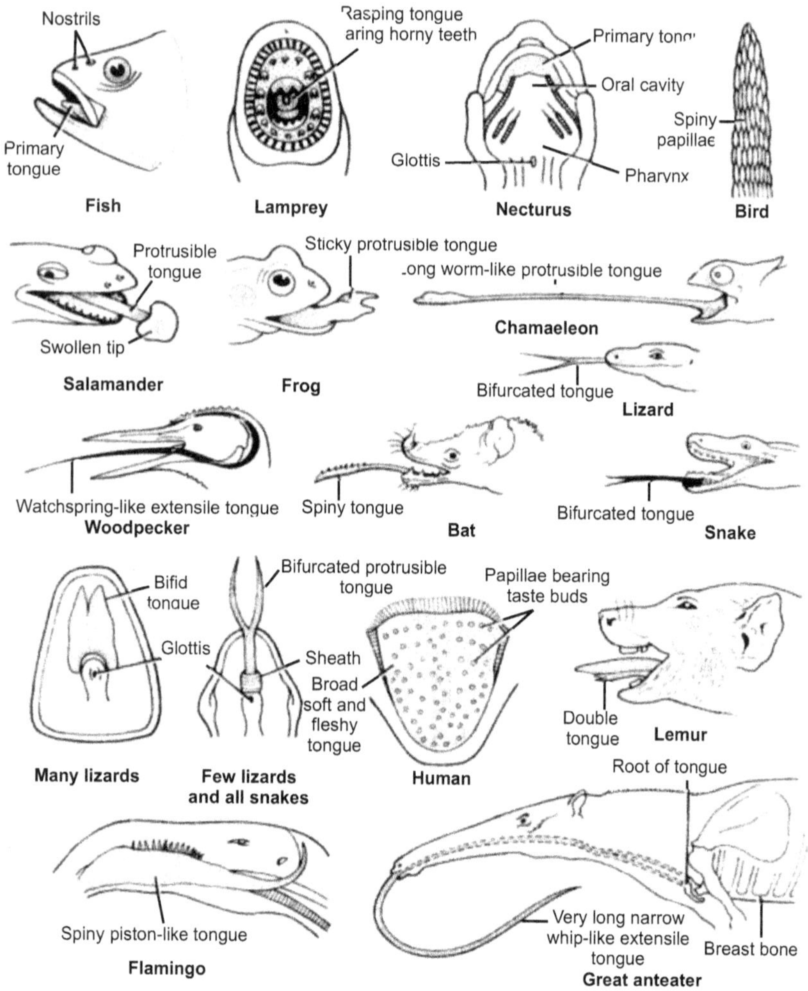

Fig. 3.3: Different types of tongues in some Vertebrates

The amniote tongue has voluntary muscles. It receives the hypoglossal nerve and has glands and taste buds. In birds, tongue is slender and has a horny covering. In some birds tongue is immobile, while in some birds it is long, protractile and often used for capturing the food. In most mammals, except whales, the tongue is highly developed and show movement. Tongue forms a median fold, called frenulum which joins the tongue to the floor of mouth. There are four types of papillae on the upper surface. They are filiform, fungiform, foliate and circumvallate.

(3) Oral glands: There are two kinds of integumentary multicellular glands opening into the buccal cavity. They are mucous glands and enzymatic glands. Fishes and aquatic amphibians have only mucous glands. Reptiles have glands in groups, such as palatine, lingual, sublingual and labial glands. In *snakes*, poisonous snakes the upper labial glands are modified to secrete venom. Birds have sublingual glands and a gland in the angle of the mouth. Mammals have many mucous glands besides true and enlarged salivary glands which are enzymatic. They are parotid, sub-lingual, sub-maxillary and infra orbital salivary glands secreting mucin and ptyalin.

(4) Adenohypophysis: The anterior lobe of pituitary gland develops as a dorsal evagination of stomodaeum, called Rathke's pouch which constricts off to form the anterior and middle lobe of pituitary gland (adenohypophysis). The posterior lobe of pituitary or neurohypophysis is the ventral evagination of diencephalon called infundibulum. Thus it is nervous part.

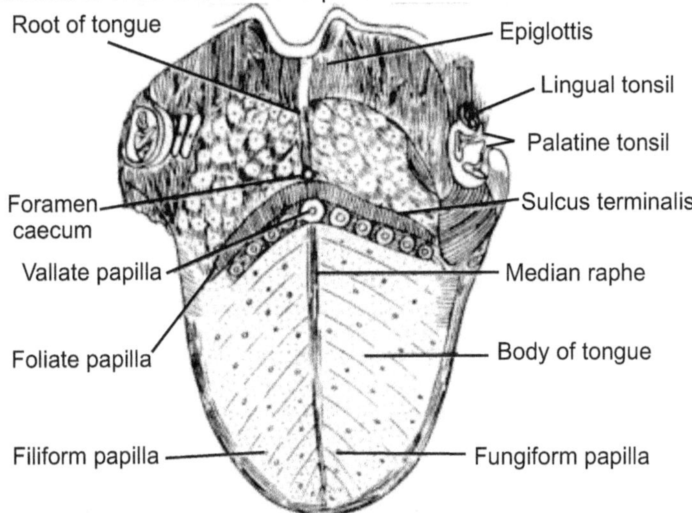

Fig. 3.4: Dorsal view of human tongue

(5) Pharynx: Region of foregut between oral cavity and oesophagus is called pharynx which is lined by endoderm. It is a common passage serving both for the digestion and respiration. It is used as a passage for food from buccal cavity to the oesophagus. Its muscles initiate swallowing. In fishes, the pharynx is large and laterally perforated for gill slits, while in tetrapoda, it is short and bears openings of nostrils. From the wall of pharynx in embryo are derived spiracle, gill clefts, lungs, air bladder, tonsils and endocrine glands such as thymus, thyroid and parathyroids.

(6) Oesophagus: It is a simple, muscular, distensible tube connecting pharynx with stomach. It is short in neckless vertebrates (fishes and amphibians), but longer in amniotes, reaching extreme in birds, giraffe etc. It may be lined internally with finger like fleshy papillae (elasmobranchs), horny papillae (marine turtles) or longitudinal folds. In grain feeding birds (pigeon), oesophagus form, a paired or unpaired membranous sac, or crop modified for storage of food. In pigeons of both sexes, epithelial lining of crop undergoes fatty degeneration controlled by a pituitary hormone, prolactin, forming pigeons milk which is fed to nestling. Oesophagus has no serous coat as it lies outside coelom. In mammals when it passes diaphragm it has serous lining. Food bolus passes down oesophagus into stomach by muscular wave of contraction and relaxation called *peristalsis* Oesophagus exhibit difference from rest part of the alimentary canal. The important differences are it has no visceral peritoneum living but outer covering is tunica adventitia. Muscle fibres of the anterior part of the oesophagus are stripped, middle part is both striped and non-striped and posterior part is only unstriped muscles. But ruminants all along their oesophagus have stripped and voluntary muscles. Internal mucosal living is of stratified squamous epithelial cells.

(7) Stomach: It is the sac like enlargement in the digestive tract between oesophagus and intestine called stomach. It is useful for storage and maceration of solid food and preliminary stages of digestion. There is no true stomach in protochordates, cyclostomes, lung fishes and primitive telecosts. When there are gastric glands present it is called true stomach. The well developed stomach is present in elasmobranches and tetrapods. The stomach shows anterior and pyloric ends. It terminates at a pyloric valve. In fishes, several pyloric caecae are present in between pylorus and duodenum. Stomach is a straight in vertebrates and remains straight in lower vertebrates throughout the life. It is longer and spindle shaped in *Proteus, Necturus,* snakes and many lizards. It is curved tube in turtles and tortoises. It is J-shaped or U-shaped in elasmobranches, seal and man. In *Polypterus,* stomach appeared like a blind pouch due to fusion of cardiac and pyloric limbs. In crocodiles and seed eating birds, stomach is divisible into an anterior thin walled, proventriculus with gastric glands, followed by thick walled highly muscular *gizzard* or *ventriculus.* The gizzard has tough, horny lining and contains small stone pieces or pebbles called *gastrolith* which helps in grinding the food.

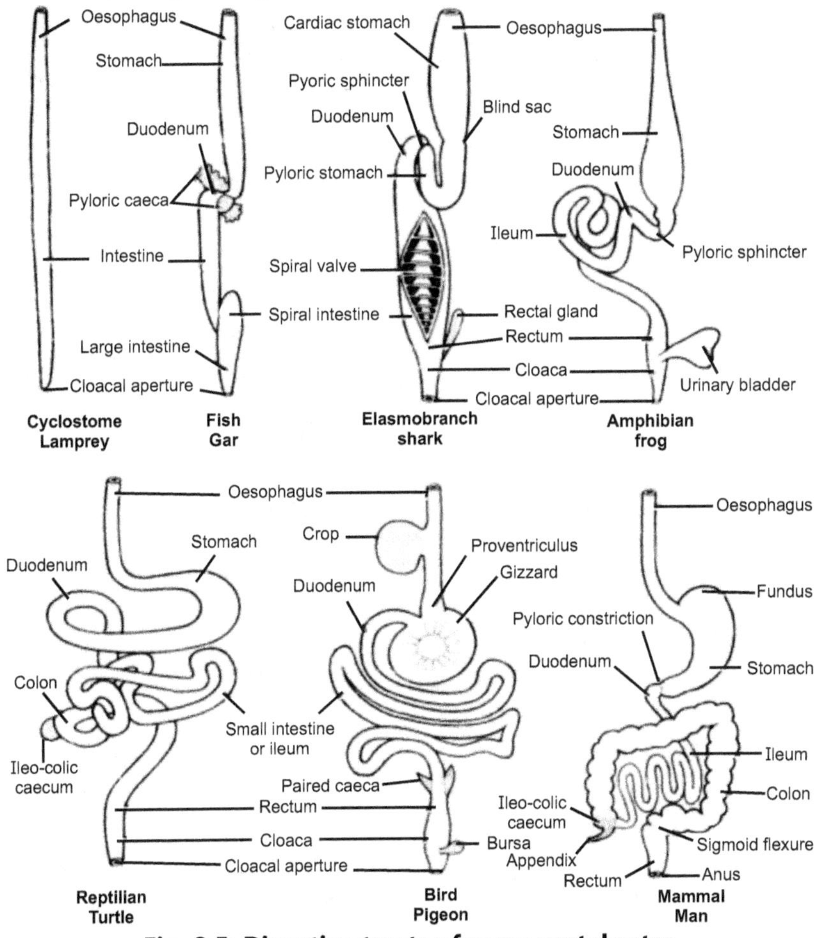

Fig. 3.5: Digestive tracts of some vertebrates

There are great modifications found in the mammalian stomach. It may be simple or divided into cardiac, fundic and pyloric regions, each region with characteristic gastric glands. In ruminants (cow, sheep) stomach had four chambers or compartments. Of these, the first three chambers (rumen, reticulum and omasum) are considered as modifications of oesophagus and serve as reservoirs of food. Only the last chamber (abomasum or rennet) represents true stomach containing gastric glands containing usual parts cardiac, fundic and pyloric. In camel, omasum is absent and pouch like water cells projecting from runen and reticulum probably helps in digestion but do not serve for the storage of water as generally believed. A pyloric enlargement in blood sucking bats serves to store blood. True stomach is lacking in monotemes only sac like stomach with no glands is present.

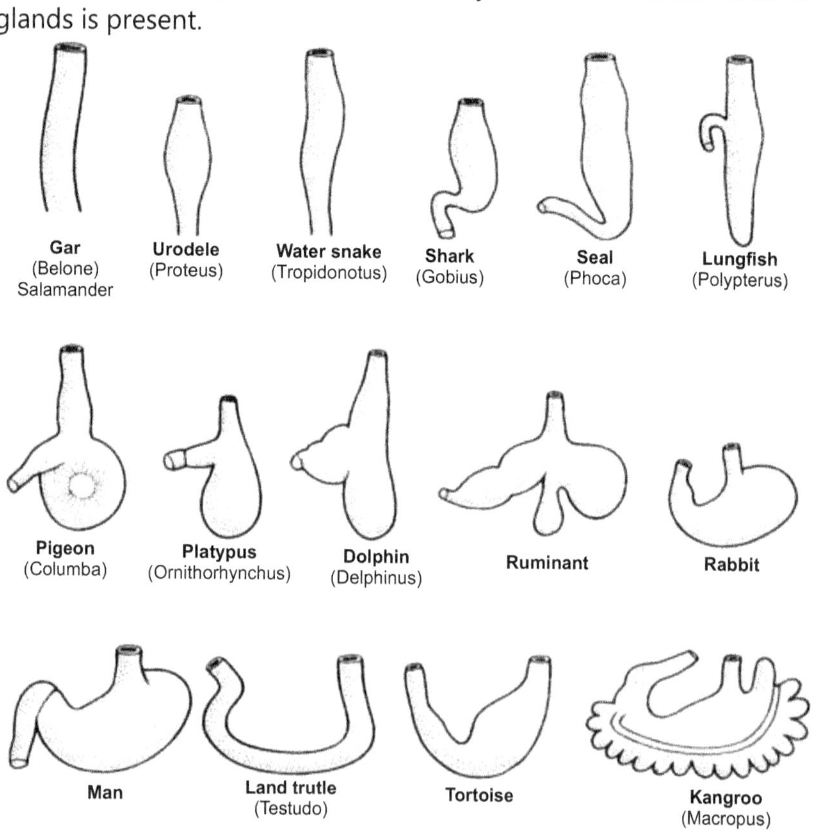

Fig. 3.6: Different shapes of vertebrate stomachs including human

(8) Small intestine: Next to stomach small intestine is the part of digestive tract, in which digestion and absorption of food takes place. It is very important part of digestive tract and undergoes several modifications in vertebrates. In cyclostomes, dog fish, bony fishes intestine is a short, straight and wide tube. Its lumen contains a typhlosole or spiral valve which helps for more absorptive area. There is no spiral valve in intestine of telecost and tetrapod; but it is greatly elongated, coiled and further differentiated into anterior small intestine and posterior large intestine. Small intestine is the main site of digestion and absorption. Large number of villi are present which increase the absorptive area. Many bony fishes have one to several pyloric caeca arising from small intestine. There are large number of digestive glands present in small intestine. They are called crypts of Lieberkuhn and secrete mucus and group of enzymes called *saccus entericus*. First part of intestine is called duodenum and it has Brunner's glands in submucosa and secrete hormones for stimulating pancreas and gall bladder to release their juices.

Duodenum is followed by remaining small intestine called *ileum*, which is narrow greatly elongated and much coiled. Only in mammals duodenum is divided into anterior *jejunum* and posterior *ileum*. Nodules of lymphoid tissues called *Peyer's patches* found in ileum.

(9) Large intestine: It is wider than small intestine in most of the fishes and amphibians. It is straight, short and leads into a posterior terminal chamber called *cloca*. Cloaca also receives urinary and genital ducts and open outside through *cloaca aperture*. In reptiles, birds and mammals, large intestine is longer and divided into a proximal *colon* and a distal *rectum*, the latter ending into colaca. All mammals except the monotremes and many bony fishes, lack a cloaca. There rectum opens directly to outside through *anus*. While the urinary and genital ducts also open separately. Rectum of mammals is derived by partitioning of embryonic cloaca and therefore, it is not homologus with the rectum of other vertebrates. In tetrapods, ileocolic spincter is present at the junction of small and large intestines, but absent in fishes. It prevents bacteria in colon from entering ileum. In amniotes, at the ileocolic junction is found an ileocolic caecum, usually two in birds. This contains cellulose digesting bacteria and it is very or even coiled, in herbivorous mammals as rabbit or horse feeding on cellulose. Man, monkeys and apes have a small caecum, bearing a vestigial *vermiform appendix*. The rectal gland of elasmobranches is a caecum that secretes sodium chloride. A blind pouch called *bursa fabrici* of lymphatic tissue arising from dorsal wall of proctodaeum in young birds but atrophies in the adults.

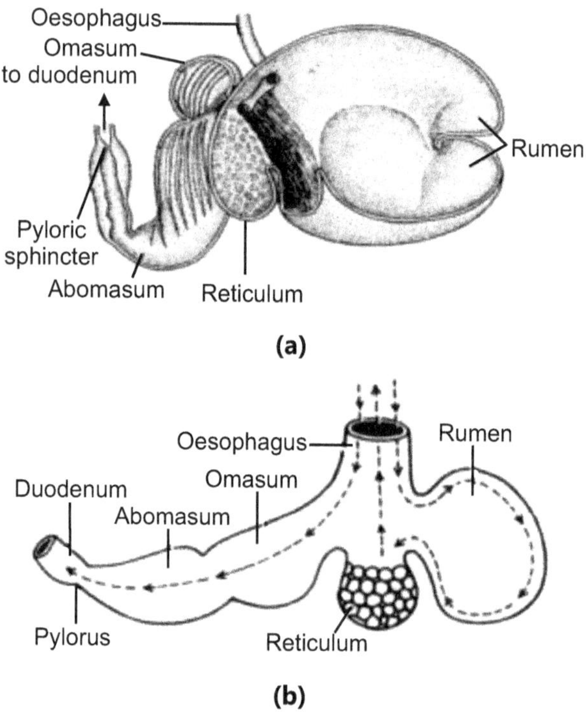

Fig. 3.7: Ruminant Stomach (a) and (b)

3.2 Digestive Glands

   (1) Liver: The liver arises as a single or double outgrowth from the ventral wall of the embryonic archenteron. The outgrowth forms a hollow hepatic diverticulum, which becomes anterior part, which ploriferates to become the liver and its bile ducts. The posterior part give rise to gall bladder and cystic duct. The bile ducts join to form a hepatic duct which unites with the cystic duct to form a common bile duct. The liver is the largest lobed gland in the body, suspended by a double layer of peritoneum from the transverse septum.

The bile juice is stored in a gall bladder. It lies in the liver and drains into the duodenum through common bile duct formed by the union of cystic duct and hepatic duct. A gall bladder is lacking in many birds and mammals.

A liver is present in all vertebrates. In cyclostomes, it small, single lobed and two lobed in hagfishes. It is bilobed in elasmobranches two or three lobed in bony fishes, amphibians, reptiles and birds and many lobed in mammals. Liver is long, narrow and cylindrical in fishes urodels and snakes. It is short, broad and flattened in birds and mammals. A gall bladder and bile duct are present in larval cyclostomes but they are absent in the adult. Fishes, amphibians and reptiles generally have a gall bladder, but it is lacking in many birds. Most mammals possess a gall bladder but it is absent in cetacea and ungulata.

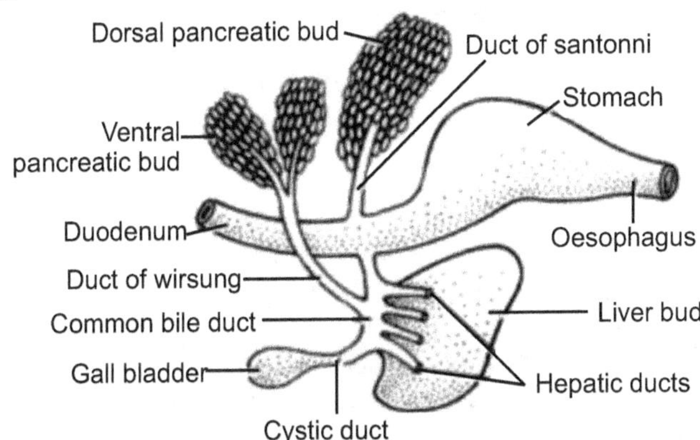

Fig. 3.8: Embryonic development of liver, gall bladder and pancreas

The liver secretes a watery, alkaline bile but have no enzymes. It neutralizes the acidity of food entering the duodenum. It aids in the digestion of fats.

(2) Pancreas: It is formed from the endoderm of embryonic archenteron. The pancreatic duct open into the duodenum either separately or after uniting with the common bile duct. Pancreas is duplex gland i.e. exocrine and endocrine gland, bound together by delicate strands of connective tissue. The exocrine part secretes digestive enzymes which are poured into the duodenum through pancreatic ducts. The endocrine part called Islets of Langerhans secretes hormones such as insulin and glucagon.

A pancreas is present in all vertebrates. In lampreys, bony fishes, lung fishes and lower tetrapods, it is a diffused organ embedded in the liver, mesenteries and intestinal wall. Hagfishes have a small pancreas. Elasmobranches have bilobed pancreas. In higher tetrapods it is generally a compact gland. One or two pancreatic ducts open into the duodenum.

Points to Remember

- Ectodermal invagination of anterior part is called stomodaeum and posterior called proctodaeum.
- Digestive tract has the following parts like mouth, buccal cavity, pharynx, oesophagus, stomach, small intestine, large intestine and cloaca.
- Mouth opening is leading into buccal cavity.
- Buccal cavity is space between lips and jaws.
- Different types of tongue is found in vertebrates.
- There are epidermal and true teeth in vertebrates.
- Stomach is not present in cyclostomes, lungfishes.
- Stomach is divisible into cardiac, fundic and pyloric regions.
- Ruminants stomach has four chambers, a rumen, reticulum, omasum and abomasum.
- Small intestine is long, narrow and coiled.
- Large intestine is with larger diameter than small intestine.
- Liver, pancreas are the important digestive glands.
- Liver secretes bile juice and it is stored in gall bladder.
- Pancreas is exocrine and endocrine gland.
- Pancreas secrete pancreatic digestive juice and is endocrine parts secretes hormone insulin and glucagon.

Exercise

1. Give a comparative account of digestive system of lizard and bird.
2. Compare the digestive system of bird with rabbit and give reasons for their differences.
3. Give a comparative account of stomach in different vertebrates.
4. Give a brief account of digestive glands.
5. Write short notes on.
   (a) Ruminant stomach   (b) Pancreas
   (c) Liver   (d) Gizzard   (e) Colon
   (f) Glands of digestive glands

Chapter 4...

Respiratory System

Contents ...

4.1 Introduction

There are different respiratory organs found in vertebrates. In frogs and salamanders skin may play role of respiratory organ; while in some fishes and aquatic turtles the vascular rectum or cloaca acts as a respiratory organ. However, there are two main types of true respiratory organs: gills for aquatic respiration and lungs for aerial respiration. In some cases, both the organs are present in some animals. Accessory respiratory organs are also present in some vertebrates. They are provided rich blood supply with thin capillaries. Moist epithelium should cover the blood vessels which are brought into close contact with the environment like water or air. Exchange of oxygen and carbon dioxide occurs at two places i.e. respiratory organs and in tissues. Internal or tissue respiration occurs between blood and tissue cells of the body. During external respiration gaseous exchange takes place between blood and external environment (e.g. in aerial respiration within lungs and in aquatic respiration within water and gill surface).

In case of lower vertebrates the respiratory organs are not connected to the olfactory organs but in air breathing vertebrates, there is close association between two. In cartilaginous fishes there is direct connection between olfactory and respiratory organs in which internal nares open from nasal cavities into buccal cavity whereas in tetrapoda the air enters through the nasal cavities into buccal cavity and then into the lungs.

4.2 Skin

Skin is also respiratory organ in some vertebrate animals, like fishes and frogs. This type of respiration is called skin or cutaneous respiration. Some fishes are able to survive outside water. The common eel, **Anguilla** can travel by wriggling on damp grass though it has no special respiratory organs, but it has vascular areas in the skin by which it can breathe both in water and on land. Secondly, the opening of the operculum is small and rounded so that the eel can retain water in the branchia chamber and journey on land. In amphibians also, the moist skin is highly vascular. Lungless salamanders respire only trough skin. Their larvae loose gills at metamorphosis and lungs do not develop in adults. African male hairy frog have vascular hairy cutaneous outgrowths which act as respiratory surface. Vascular caudal fin of mud skipper acts as respiratory organ during submergence. In the mud skipper of the Indian and Pacific Oceans the caudal fin is highly vascular, the head and trunk of the fish project above water when it perches on a rock, only the caudal fin remains submerged and acts as a respiratory organ.

The frog is amphibious animal, passes most of the time of its life in water. During this period the skin only serves as an organ of respiration for gaseous exchange. Similarly when frog undergoes summer sleep (aestivation) and winter sleep (hibernation), the skin plays an important role of respiration. The skin of frog is very much suited for respiratory function as it is very thin and richly supplied with blood capillaries and remains moist with the water and also mucus secreted by mucous glands. During gaseous exchange the oxygen first dissolves in the moisture present over the body and diffuses into the blood circulating in the blood capillaries, while the resultant carbon dioxide passes out from the blood into the surrounding medium by diffusion. In the cutaneous respiration, no movements are needed because skin always remains exposed to air or water.

4.3 Gills

Gills are the important respiratory organs found in fishes and amphibians. Gills also play important role of salt elimination in marine teleosts. On the basis of position gills are of two types: external gills and internal gills.

(1) External Gills: In most of the amphibian larvae show the fuft. of filaments originated from the visceral arches called external gills and they are respiratory in function. They are ectodermal in origin. In tailed amphibians, external gills and gill slits are retained throughout the life. In case of some tailed and tailless amphibians the external gills are lost during the metamorphosis and hence they are called *larval gills*. Though in embryo there are five pairs of pharyngeal pouches arise, out of which only 2, 3 and 4th perforate and open to the exterior. The 3rd, 4th, and 5th visceral arches bear external gills. The external gills are exposed in the water and gaseous exchange occurs through their surface epithelium. Later an operculum arises and covers these gill clefts and gills externally in tadpoles so that the gills become enclosed in an operculum chamber lined with ectoderm.

These external gills soon degenerate and a new set of gills called *internal gills* originate from the same visceral arches. External gills are rare in fishes, but found in some larval forms of lamprays, in polypterus has one pair of external gills. In case of *Lepidosiren* (Dipnoi) there are four pairs of filamentous external gills. The external gills disappear in fishes in adult stage. The gills may be pectinate, bipinnate, dendric or leaf like. The gill has a narrow central axis bearing double row of filaments. These are richly vascularised by aortic arches.

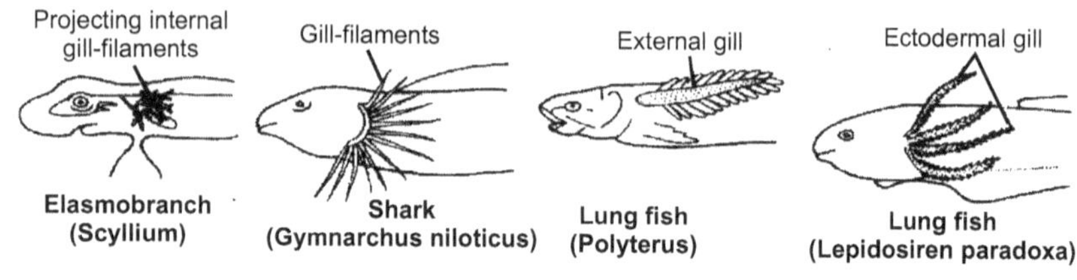

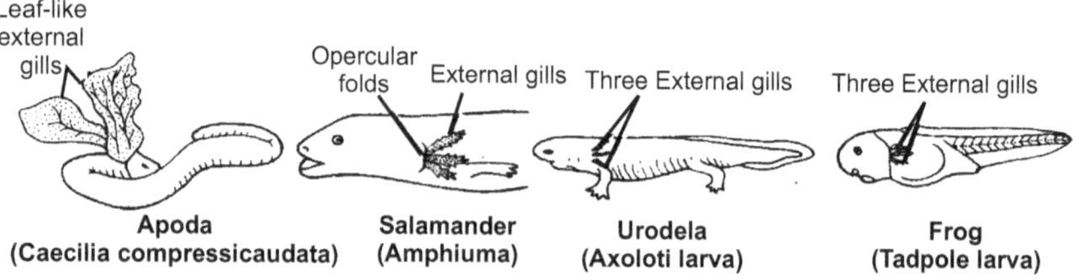

Fig. 4.1: Larval external gills of fishes and amphibians

(2) Internal or true gills: Internal gills are characteristic of fishes. These are situated in the gill slits and attached to the visceral arches. In amniotes embryonic pharyngeal pouches do not communicate to the outside through gills slits in the adults. Gills are absent in them.

Gill slits: Gill slits are one of the most fundamental traits of the chordata. In the embryo, the pharyngeal cavity is connected to the outside by a series of lateral openings known as pharyngeal clefts or simply gill slits. These persist in the adult state of protochordates, cyclostomes, fishes and certain amphibians, but become reduced, abolished or modified in higher vertebrates. The number of gill slits varies in different chordates, 140 in Amphioxus, 6-14 pairs in Cyclostomes, 5 pairs in Elasmobranchs, 6 pairs in *Hexanchus,* 7 pairs in Heptanchus, 4 pairs in *Chimeras,* 5 pairs in most of the bony fishes and 4 pairs in some teleosts. The gill slits are separated from one another by partitions called visceral or gill arches. The arches are supported by skeletal structures of splanchnocranium, together forming the visceral skeleton.

Structure of true gill: These gills develop on the walls of gill pouches or gill arches. A gill is formed of two rows of a series *gill filaments* or *gill lamellae* develop from the epithelium covering the *inter branchial septum* on both sides. This septum extends outward from the cartilaginous arch. A single row of gill filaments on one side of interbranchial septum forms a half gill called a *hemibranch* or *demibranch*. The complete gill or holobranch is formed with interbranchial septum with two hemibranchs (anterior and posterior). These are richly supplied with blood vessels associated with aortic arches so that carbon dioxide in the blood may be exchanged for dissolved oxygen in the water.

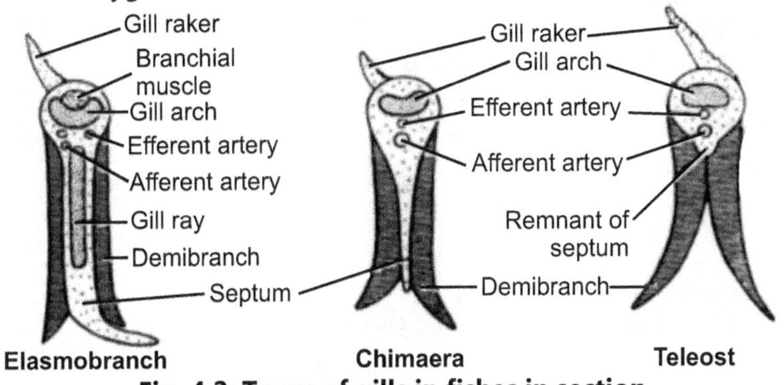

Fig. 4.2: Types of gills in fishes in section

Sharks and rays have generalised structure of gills. In higher bony fishes the interbranchial septum is lacking so that the hemibranchs on the anterior and posterior part of each branchial arch are no longer separated from one another. Furthermore, the gill apertures also no longer open separately to the outside. Instead, the gills are enclosed in a single chamber and covered externally by a large bony operculum which opens and closes posteriorly to permit water to pass to the outside. In most bony fishes there are four pairs of functional gills.

Spiracles: In sharks and rays an anterior pair of non-respiratory opening, one on either side between mandibular and hyoid arches are called spiracles or pseudobranchs. These are internally closed. These become closed or lost in lung fishes and bony fishes.

4.4 Lungs

Most of amphibians and all amniotes breathe by means of lungs. Lungfishes also possess lungs. From the ventral wall of the pharynx during embryonic development a hallow outpushing called lung premordium arises. It grows backwards and divides into two right and left lungbuds. The undivided proximal portion develops into *trachea* and *larynx* and opens into pharynx by glottis. Later lung buds grow posteriorly into coelom and branch repeatedly and get covered by mesoderm. Thus, each lung has an endodermal lining and an outer visceral peritoneum and in between two mesodermal mesenchyme having blood and lymph vessels, nerves and smooth muscle fibres and connective tissue. Inner endodermal epithelium of lungs is raised into a network of ridges to increase the vascularised surface exposed to the action of air.

In lower forms, the lungs are hollow bags, but in higher forms the ridges increase in number and unite with one another across the lumen of the lung to convert it into a solid but spongy structure with innumerable air spaces. In mammals, the internal surface area of lungs is 30% more than the external surface area of the body. The original duct of the lung sac connecting the pharynx to the lungs becomes a *trachea* in most. There is no trachea in anurans. In many tetrapoda anterior end of the trachea becomes modified into a *larynx* or *sound box*, which opens into the pharynx by a glottis. The trachea divides into two bronchi, each of which enters a lung. Bronchi divide to form an immense system of bronchioles carrying air into minute bags or alveoli. The alveoli have very thin walls invested with blood capillaries, an exchange of gases occurs in the alveoli.

Trachea: It is very short or absent in anurans. It is merged with the larynx to form laryngotracheal chamber. Many tailed amphibians possess short trachea supported by trachea. Trachea is simple in reptiles may be long in long necked reptiles such as turtles, trachea is long and convoluted. In birds trachea is long. Trachea in mammals is variable and tracheal rings are usually incomplete on upper side.

Lungs: In *polypterus* paired ventral lungs are present which enable these to survive during drought conditions. In all the living lungs fishes, the lung is dorsal to the gut connected by a tube to the ventral side of the oesophagus. In *Protoptenus* (African) and *Lepidosiren* it is biolobed and unpaired in Australian lung fish. In most bony fishes primitive lung has modified into a gas or swim bladder or hydrostatic organ. It is connected to the oesophagus by dorsal connection.

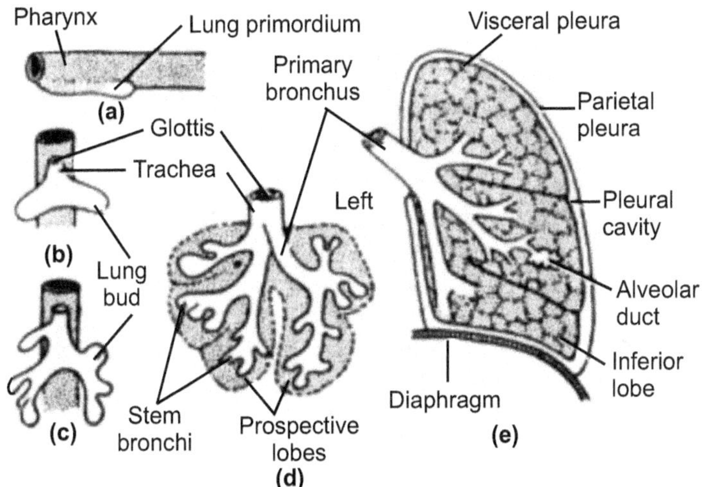

Fig. 4.3: Development of vertebrate lung in embryo

In amphibians the lungs are simple, sac like structures with a central large cavity. In frogs and toads the inner walls contain numerous folds lined with alveoli to increase the respiratory surface.

In reptiles, lungs are more complex with increasing number of internal chambers and alveoli. In some lizards, one lung is considerably large than other; and in snake the left lung is reduced or even absent in some species. Crocodilians have the lungs like mammals. In some lizards, bronchi are subdivided into primary, secondary and tertiary bronchi.

In birds, lungs are small and do not show great expansion. They are connected with nine air sacs which are situated in different parts of the body. Air sacs have no respiratory epithelium but they serve as reservoirs. Air passes through a bronchial circuit into air sacs and then returns, generally by a separate set of bronchi, to the air capillaries in the lungs.

In case of mammals, respiratory system is much less complicated than birds. The primary bronchi after entering the lung into secondary bronchi which divide into smaller and smaller bronchioles finally terminating in tiny alveoli where exchange of gases takes place. In majority of mammals, lungs are subdivided externally into lobes; i.e. left lung has two lobes and right lung has three lobes in man and four lobes in rabbit. In whales, sirenians elephants, hyrax lungs are simple and without lobes. Right lung is lobulated in Monotremes and rats. In Siren lungs are elongated.

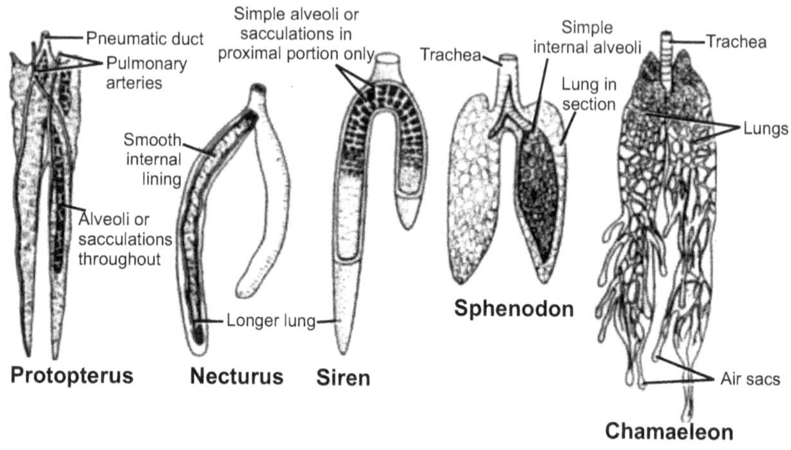

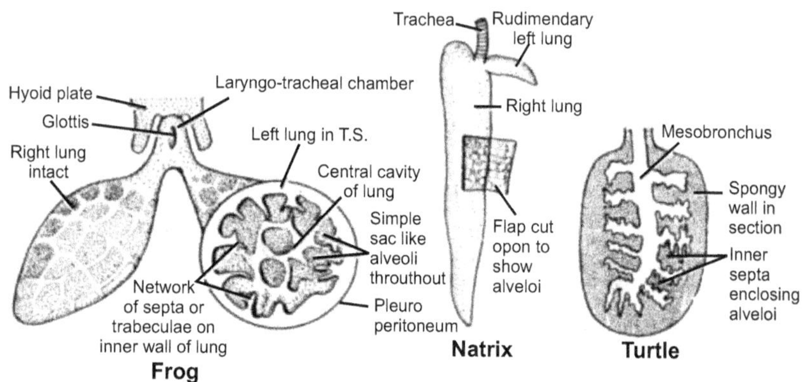

Fig. 4.4: Different types of vertebrate lungs

Mechanism of Respiration: In fishes and amphibians, the mechanism of respiration is same. The floor of the buccal cavity is lowered and water (fishes) or air (amphibians) is taken in the mouth is closed and floor of the buccal cavity is raised which forces the water into gill clefts in fishes or the air into the lungs in amphibians. In amniotes air is taken in by increasing the volume of the lungs by an expansion of the thorax, this is done by movement of the ribs (and by movements of the diaphragm in mammals). In turtles, where ribs are fixed to the carapace, the volume of the lungs is increased by movement of the neck and limbs.

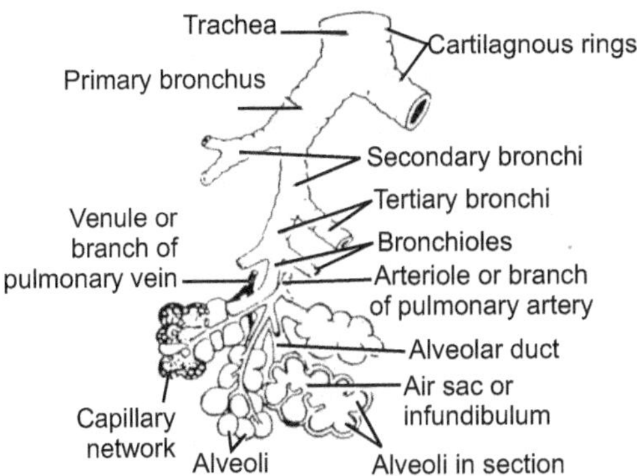

Fig. 4.5: mammalian lung.
Branching of a bronchus into terminal alveoli

4.5 Accessory Respiratory Organs
4.5.1 Air Sacs

These are found in birds only. They are dilations of the bronchi. They are large, thin walled membranous, non-muscular and non-vascular structures. They do not increase respiratory surface area and lie in viscera and in bones. They communicate with bronchi. There are nine air sacs and named according to their position in the body. Interclavicular, cervical, anterior thoracic, posterior thoracic, abdominal are the different types of air sacs.

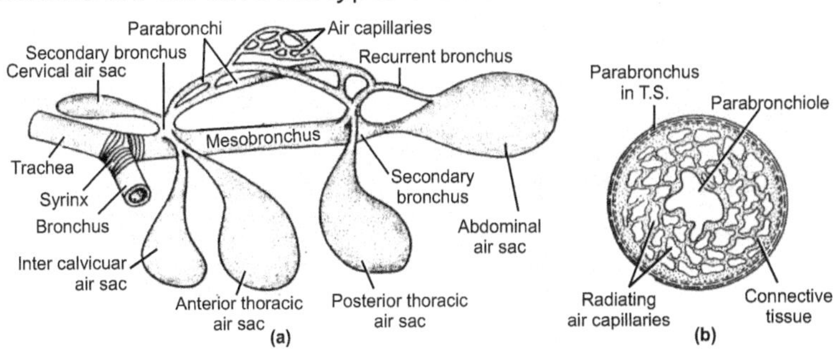

(a) Scheme of architecture of a lung and air sacs in a bird

(b) Section of a parabronchus showing radiating infundibular or air capillaries

Fig. 4.6: Air sacs in birds

The air sacs are not respiratory organs but they help in respiration. They act as bellows forcing their air into the lungs for ventilation at each expiration to completely renew the air in the lungs, thus there is no 'dead space' of unrespired air in the lungs. But it is claimed that anterior air sacs are expiratory and they are more active during flight. The posterior air sacs are inspiratory. They are more active when the bird is not flying.

Besides this, they are useful for lightness, temperature regulation, cardiac movement and flight.

4.5.2 Swim Bladders

Another important structure serving as a lung in some lower fishes is the swim bladder or air bladder. The Indian climbing perch *Anabas* has special air chambers above the gills, where three concentrically folded bony laminae called labyrinthi form organs are developed from the first epibranchial bone on each side. This covering vascular mucous membrane brings about respiration. *Anabas* is so dependent on air that even in water it comes to the surface to gulp air and it is asphyxiated if prevented from doing so. It can survive for long periods on land and makes excursion by means of its many long spines on openculum and ventral fin. In *ophiocephalus* there is an accessory branchial cavity on each side above the gills.

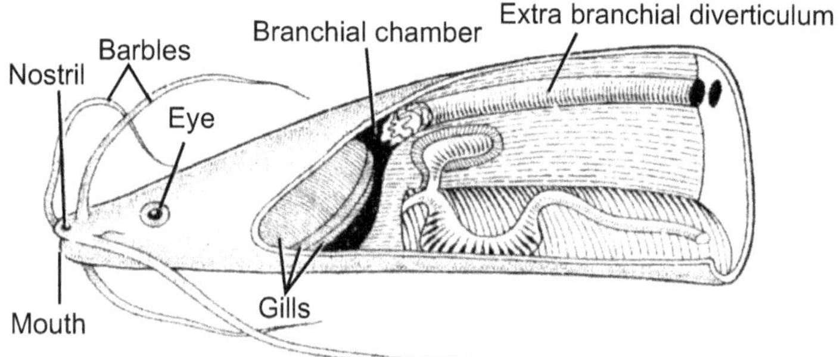

(a) Accessory respiratory organs of *Saccobranchus*

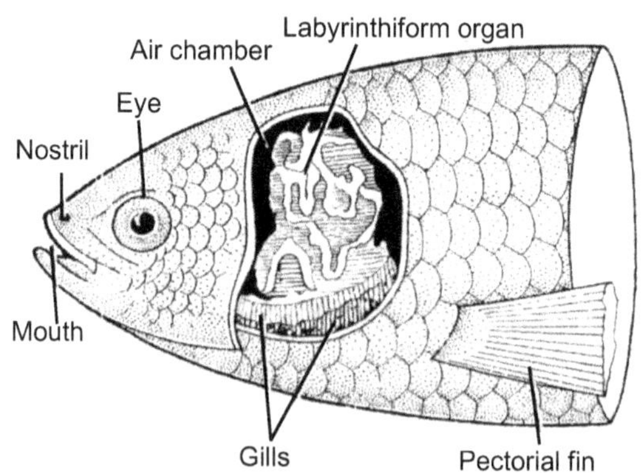

(b) Accessory respiratory organs of *Anabas*

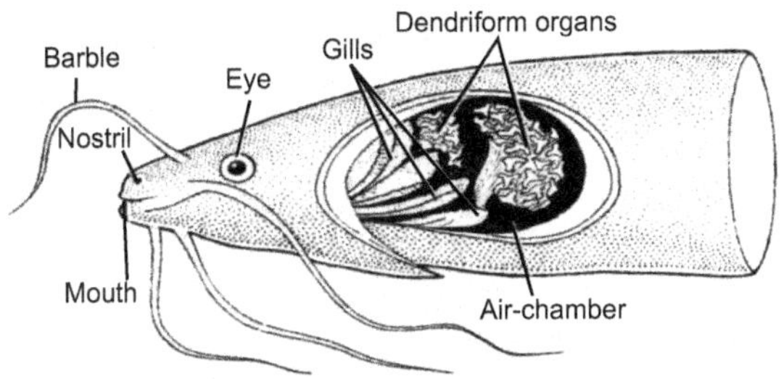

(c) Accessory respiratory organs of *Clarias*

Fig. 4.7: Diagram showing the accessory organs of (a) *Saccobranchus*, (b) *Anabas*, (c) *Clarias*

Points to Remember

- Gills are important respiratory organs in water.

- Lungs are useful for aerial respiration.

- Accessory respiratory organs are also present in some vertebrates.

- Skin, swim bladder, epithelial lining, are the accessory respiratory organs.

- Internal or true gills are characteristics of fishes.

- Gills slits persist in adult cyclostomes, fishes and certain amphibians.

- Amphibians and all amniotes breathe by means of lungs.

- In lower forms, lings are hollow bags.

- Large number of alveoli are present in the lungs of mammals.

- In *Polypterus fish* paired ventral lungs are present useful in drought conditions.

- Reptile lungs are more complex than amphibians.

- Mammalian lungs are less complicated than birds.

- Swim bladders, skin, pharyngeal diverticula are the accessory respiratory organs in fishes and amphibians.

Exercise

1. Give a comparative account of respiratory organs in vertebrates studied by you.

2. Write a detailed essay on accessory respiratory organs in vertebrates.

3. Write short notes on:

   (a) Larval gills   (b) Pseudobranchs

   (c) Lungs   (d) Air bladder

   (e) Swim bladders   (f) Structure of true gills

   (g) External gills   (h) Internal gills

   (i) Accessory respiratory organs

   (j) Cutaneous respiration.

Chapter 5...

Circulatory System

Contents ...

5.1 Evolution of Heart

Heart is an important pumping organ of the vertebrates. It is unpaired organ but its origin is bilateral. The development of heart occurs in the embryo from mesenchyme. It forms a group of endocardial cells below the pharynx. These cells become arranged to form a pair of thin endothelial tubes.

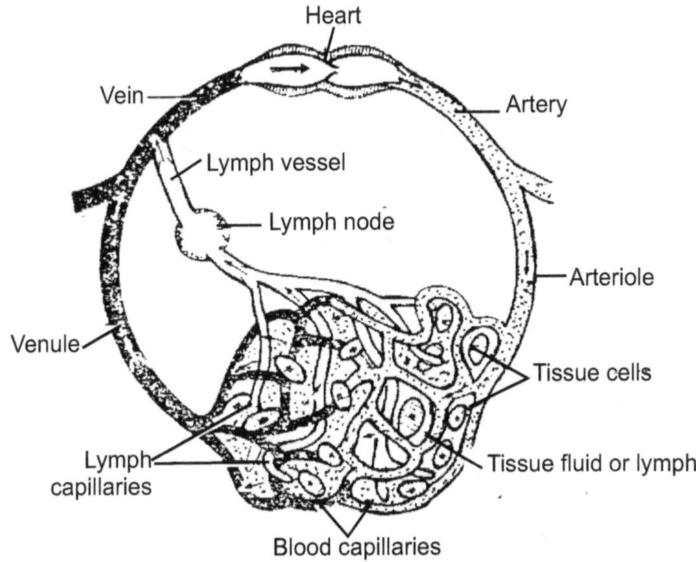

Fig. 5.1: Fundamental structure and parts of a typical mammalian circulatory system

Then, soon these two tubes fuse with each other to form a single endocardial tube which is lying longitudinally below the pharynx. In case of fishes, reptile and birds the endocardial tube is formed by coming together of two vitelline veins while in mammals endocardial tube is formed from a single endothelial tube.

Below the endoderm splanchnic mesoderm is lying which gets folded longitudinally around the endocardial tube, resulting two layered tube from the heart. The splanchnic mesoderm becomes thick and forms the muscular wall or *myocardium* of the heart. The outer thin layer is called *visceral pericardium* or *epicardium*. The endocardial tube becomes the lining of the heart called *endocardium*.

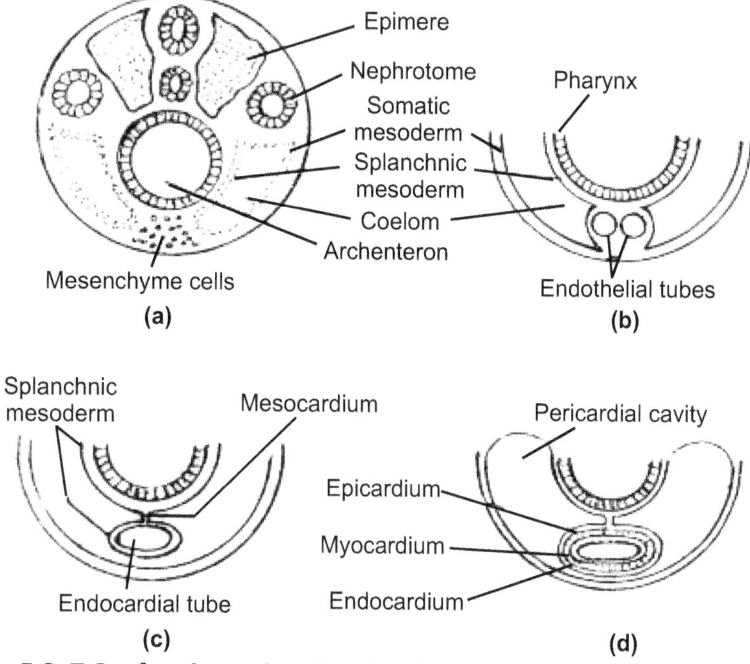

Fig. 5.2: T.S. of embryo showing development in the stages of heart

The folds of splanchnic mesoderm meet above to form a dorsal mesocardium which suspends the heart in the coelom. Soon a transverse septum is formed behind the heart which divides the coelom into two chambers. The anterior chamber is called pericardial cavity which encloses the heart. The posterior chamber is called posterior abdominal cavity. Thus, heart is a straight tube and it increases in length and becomes S-shaped because its ends are fixed. The appearance of valves, constrictions, partitions in the heart and differential thickenings of its walls form three or four chambers in the heart. The vertebrate heart is built in accordance with the basic architectural plan.

5.1.1 Comparative Study of Heart

(I) Single Chambered Heart:

This is actually a part of ventral aorta situated below the pharynx which becomes muscular and contractile. Many zoologists consider this tube as a single chambered heart. This is found in primitive chordates called cephalochordata.

In **Branchiostoma** or Amphioxus, a cephalochordate, true heart is lacking. But a part of ventral aorta functions as a heart for pumping the blood.

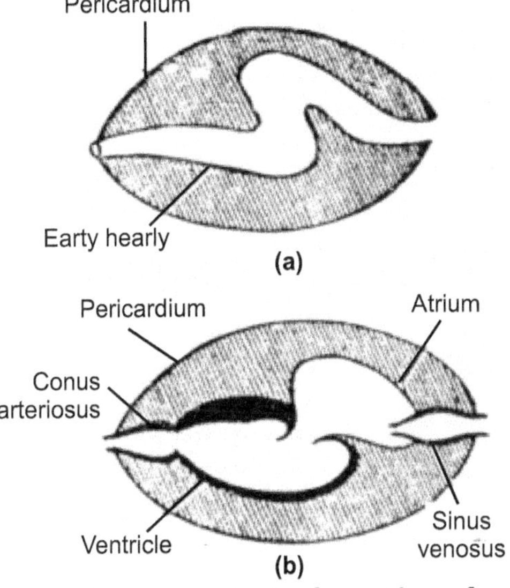

Fig. 5.3: Stages in the formation of heart

Single chambered heart found in primitive forms get modified into higher forms. Following changes occur during modifications.

(1) Chambers are formed by constriction in the cardiac tube.

(2) Further each chamber tends to divide into two separate chambers by formation of partition.

(3) There is shifting of heart from just behind head (fishes, amphibians) near gills into thoracic cavity (amniotes) with elongation of neck and development of lungs.

(II) Two-chambered Heart:

This is also called single circuit venous heart. This type of heart is found in *cyclostomes*. This is simplest type of heart and it shows thin walled *sinus venosus* which opens into slightly muscular atrium or auricle. The atrium leads into a muscular ventricle which opens into a muscular *conus arteriosus* or *bulbus cordis*. In between atrium and the ventricle the endo-cardium forms an atrio-ventricular valve. There are also valves in conus arteriosus. The blood flows in that sequence.

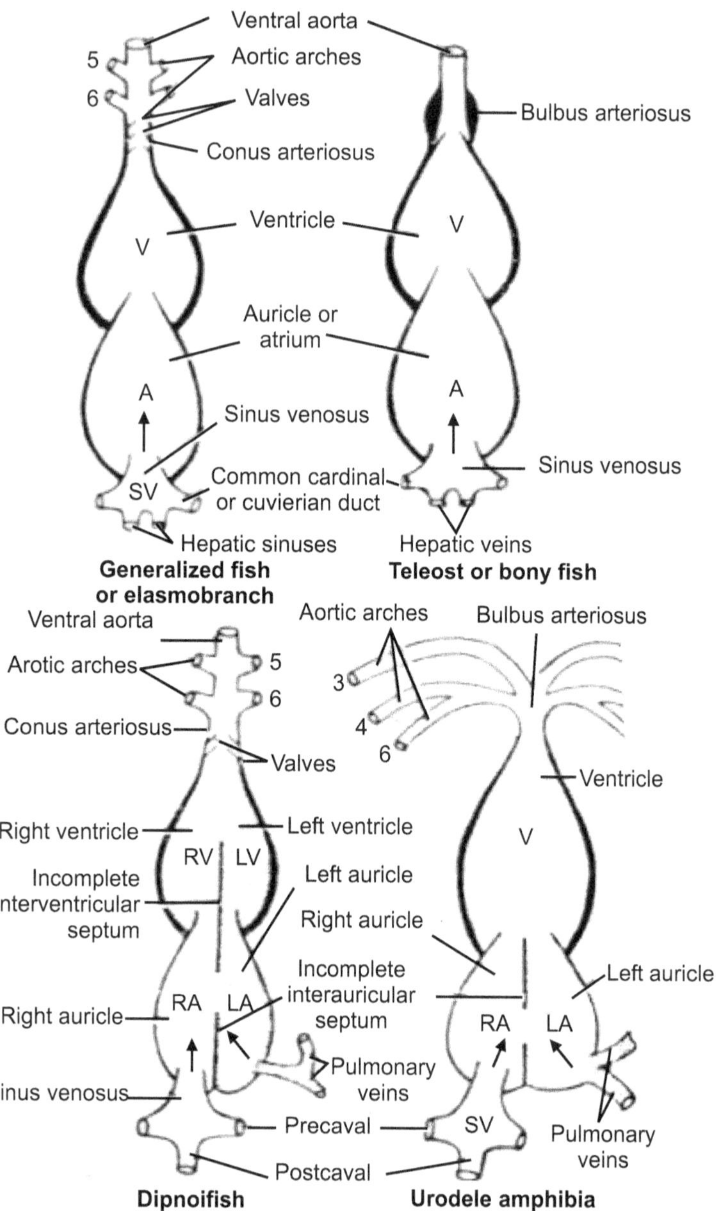

Fig. 5.4: Comparative study of hearts in vertebrates 1-6 represents aortic arches. LA - Left aurical, LV - Left ventrical, RA - Right aurical, RV - Right ventrical, SV - Sinus venous

Elasmobranchs: The heart of the cartilaginous dog fish is typical and generalised for most fishes. The heart is muscular, dorsoventrally bent, S-shaped tube consisting of four chambers or compartments arranged in a linear sequence. *Auricle* and *ventricle* are the true chambers and *sinus venosus* and *conus arteriosus* are accessory chambers. Therefore, the heart of the fishes is considered as two chambered heart. The sinus venosus is thin walled which receives venous blood of body through larger veins common cordinal and hepatic. Sinus venosus serves as a reservoir and opens anteriorly into atrium through since atrial aperture is guarded by a pair of valves. Atrium is large, thin walled, elastic and muscular chamber situated dorsal to the ventricle. Ventrally, it opens into the ventricle through an atrio-ventricular aperture guarded by a pair of valves. Ventricle has thick and muscular walls. The ventricle opens into a narrow muscular called conus arteriosus which possess many semilunar valves. The valves present in the heart prevent the back flow of blood.

The heart of fishes is encapsulated by a small pericardial cavity. This cavity is separated from general coelom by a transverse septum. The conus joins with ventral aorta in front of pericardial cavity. The transverse septum is perforated by a pair of openings through which pericardial cavity communicates with coelom in case of elasmobranchs.

Teleosts: The heart of the teleosts or bony fishes is very similar to that of elasmobranchs. The conus is comparatively very large in Cinondrosteri (**Polypterus**) and Holostei (**Lepidosteus**) with many valves. But in case of *Amia* the conus and the number of valves are considerably used. In Teleosts conus is very much reduced or even absent, because it fuses with the ventricle and bears only pair of semilunar valves. Instead, the part of ventral aorta in contact with conus becomes greatly enlarged with thick muscular walls and it is called *bulbus arteriosus.*

The fishes show small two chambered heart with a single circuit of blood circulation. All blood passing only once through heart is non-oxygenated. Before its distribution to body organs it is pumped into gills for aeration. Therefore, such heart is known as branchial or venous heart.

(III)Three-chambered Heart:

This is transitional condition and found in Dipnoi fishes, amphibians and reptiles.

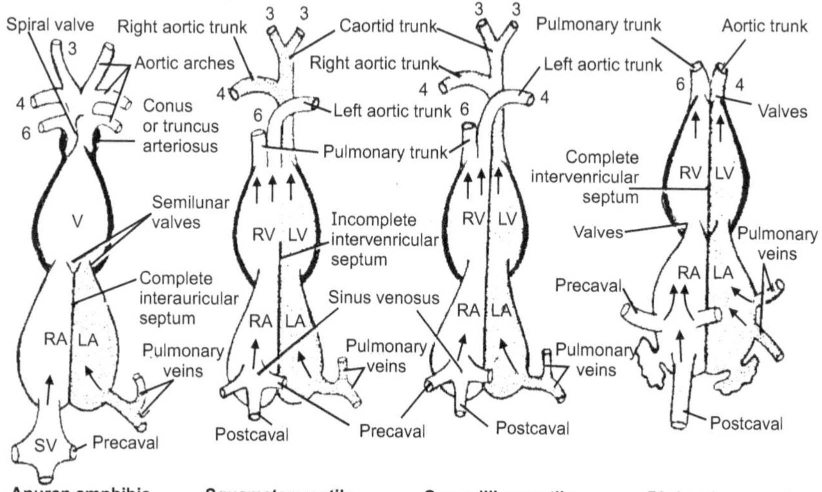

Fig. 5.5: Successive modifications of heart during evolution in different classes of vertebrates. 1-6 represents aortic arches.
LA - Left aurical, LV - Left ventrical, RA - Right aurical,
RV - Right ventrical, SV - Sinus venous

Dipnoi: Dipnoi or lung breathing fishes show the shift from aquatic or gill respiration to terrestrial or lung respiration. Therefore, the heart and aortic arches also show modifications in this group. Along with the systemic circulation there is begining of pulmonary circulation. Because of this development aerated blood from lungs or swim bladder returns directly to the heart without making circulation of the whole body. The heart of lung fishes and also most of urodel amphibians show incomplete division of atrium. The partition is called inter-auricular septum which is perforated by *foramen* ovale. Due to this partition right and left auricles are formed. This results in a mixing of oxygenated blood received lungs into left auricle and deoxygenated blood from rest of the body into right auricle. The ventricle is also divided by a partial partition or septum. The conus of lung fishes is divided by a horizontal septum into dorsal and ventral parts.

Amphibians: As compared to the heart of fishes, amphibian heart is more advanced. More modifications are seen in amphibian heart. For example, a twisting or curving results in dorsal atrium shifting anteriorly to ventricle. Similarly, sinus venosus opens into the right atrium dorsally instead of posteriorly.

The inter auricular septum is complete without foramen ovale. Therefore, complete left and right auricles are formed. The oxygenated blood comes to the left auricle and deoxygenated blood into the right auricle. This arrangement keeps the oxygenated and deoxygenated bloods separate. The deoxygenated blood from right auricles has to go the lungs for aeration and this necessitates a double circulation through the heart, once an oxygenated blood stream coming to the heart and once a deoxygenated blood stream from the great veins. This type of heart is called *pulmonary heart*. The ventricle remains single or undivided but its wall is thick and muscular. Inner side of the wall is raised into trabeculae permitting only little mixing of two bloods. In urodel amphibians, the conus is reduced and replaced by a bulbus arteriosus. In *anurans* conus or truncus arteriosus is prominent and it is divided internally by spiral valve which directs deoxygenated blood into pulmonary vessels and oxygenated blood into systemic vessels.

Reptiles: Reptiles heart is more advanced than amphibian heart, because it shows two auricles and two ventricles. In many reptiles ventricle is partially divided by an interventricular septum which reduces the mixing of oxygenated and deoxygenated blood.

Crocodiles are the first reptiles in which the septum is complete, thus making an effective four-chambered heart with two auricles and two ventricles. However, no complete separation of oxygenated and deoxygenated blood is achieved in reptiles. This is because the right and left systemic aortae, carrying arterial and venous bloods respectively, join to form the dorsal aorta in which two bloods get mixed before distribution. Further, *foramen* of *Panizza* is a small opening which connects two aortae at their base, bringing about some mixing of blood. All reptiles show sinus venosus. Conus and ventral aorta of embryo become split in the adult into three distinct trunks-pulmonary and right and left systemic.

Amphibians and reptiles represent transitional hearts because they show midway condition between two chambered heart of fishes with single circulation and four-chambered hearts of birds and mammals with double circulation and complete separation of arterial and venous bloods.

(IV) Four-chambered Hearts:

These are also called double circuit pulmonary hearts. These hearts are represented by birds and mammals.

Birds and mammals: The hearts of birds and mammals are four-chambered because the ventricle is completely divided by inter *ventricular septum*. Thus, heart has two auricles and two-ventricles. Because of formation of two ventricles the two blood streams never mix, the oxygenated blood lies in the left auricle and ventricle and deoxygenated blood in the right auricle and ventricle. This brings about a complete double circulation. Left auricle receives oxygenated blood from lungs, pour into left ventricle which pumps it to entire body through *systemic circulation*. Right auricle receives deoxygenated blood returning from body, passes to the right ventricle which pumps it to lungs for reoxygenation. This is called pulmonary circulation. Thus, there is double circulation in which there is no mixing of oxygenated and deoxygenated blood at all. Such heart is called pulmonary heart. Sinus venosus is absent because it is completely incorporated into right auricle which directly receives two pre-cavals and post-cavals. Similarly, the left auricle receives blood directly through pulmonary veins. Primitive conus arteriosus is completely replaced by a pulmonary aorta leaving right ventricle for lungs and single systemic aorta leaving the left ventricle for body. All major vessels have valves basally at the point of exit or from entry into heart.

5.2 Evolution of Aortic Arches in Vertebrates

Embryonic plan of arteries: When the heart is being formed in a vertebrate embryo a blood vessel called *ventral aorta* appears midventrally below the pharynx, it soon get connected to the conus arteriosus. The ventral aorta runs forwards and divides at its anterior end into two *external carotid* arteries into head. Ventral aorta gives off, at intervals 6 pairs of *aortic arches* running through the visceral arches. Each aortic arch consists of a ventral *afferent branchial* artery carrying venous blood to capillaries in a gill and *efferent branchial artery* taking arterial blood from the gill. All efferent branchial arteries of the same side dorsally join to the *lateral dorsal aorta* or *radix* which extended into head as the internal carotid artery. The first aortic arch is called mandibular aortic arch, the second is a hyoid aortic arch, the remaining ones are called 3rd, 4th, 5th and 6th aortic arches. The two lateral dorsal aortae unite just behind the pharynx to form a single median dorsal aorta which continues behind into tail region as *caudal artery.* From the dorsal aorta paired and unpaired arteries arise which supply various organs of the body. In an embryo with a yolk sac a pair of vitelline arteries arise from the dorsal aorta and supply the yolk sac. In embryos of amniotes a pair of umbilical or allantoic arteries arise from the dorsal aorta and go to the allantois.

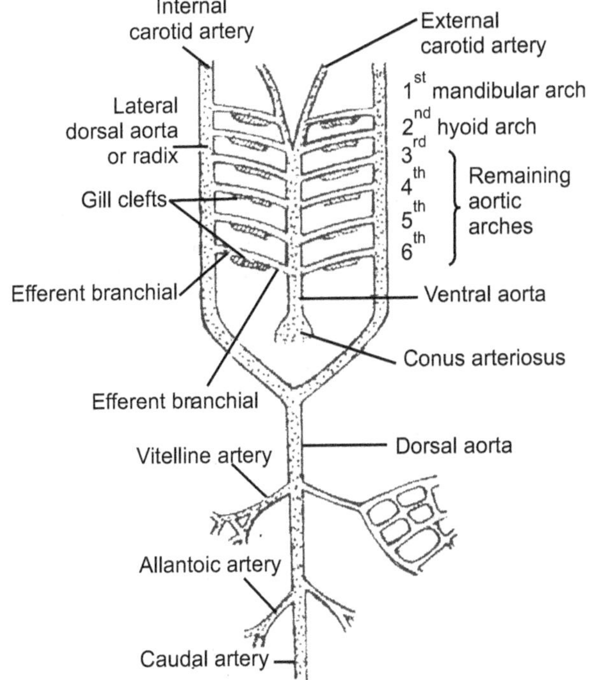

Fig. 5.6: Basic pattern of chief arterial channels of a typical vertebrate embryo

In an adult, the vitelline arteries fuse to form the main mesenteric artery, the major part of the allantoic arteries is lost, but their remnants form hypogastric or internal iliac arteries.

Although arterial system of different adult vertebrates show major differences, but it is actually built according to the same basic architectural plan seen in the vertebrate embryo. The modifications or differences are seen only because of increase complexity of heart on account of a shift from gill respiration to lung respiration.

There is a progressive reduction in the number of aortic arches in the vertebrate series.

Primitive vertebrates: In case of lower vertebrates like ***Branchiostoma*** (Amphioxus) there are nearly 60 pairs of aortic arches are present which connect ventral and dorsal aortae. In *Petromyzon* 7 pairs of aortic arches are found. Whereas in other cyclostomes the number of aortic arches vary from 6 pairs (*Myxine*) to 15 pairs (*Eptatretus*). In these forms aortic arch show two parts, an *afferent branchial artery* taking blood from the ventral aorta to a gill, where it forms capillaries and an efferent branchial artery arising from gill capillaries and taking oxygenated blood to the lateral dorsal aorta.

Fishes: The primitive elasmobranchs (*Heptanchus*) show 7 pairs of aortic arches. Most of the fish embryos present primitive plan with 6 or more pairs of aortic arches, each passing through gill. But in adults the mandibular aortic arch is modified or lost so that there are only five aortic arches present in the adult. In most sharks only five pairs of aortic arches are (II, III, IV, V and VI) functional. The first gill slit called spiracle which is non-functional as a gill. In case of bony fishes the first and second arches disappear and therefore only four pairs III, IV, V and VI) of arches remain functional. In lung fishes or Dipnoi, gills are poorly developed but *pulmonary artery* arises on each side from the sixth aortic arch or dorsal aorta and supplies blood to the air bladder or lung. In case of *Protopteus* the III and IV embryonic arches are uninterruped by gill capillaries.

In elasmobranchs and Dipnoi, each arch forms one afferent and two efferent arteries (formed by splitting) in each gill. But in case of teleosts or bony fishes, each gill has one afferent and one efferent artery.

Amphibians: In tetrapoda, there is further reduction in the number of aortic arches and they do not break up into afferent and efferent parts because true internal gills are absent. I and II arches totally disappear in all tetrapods. In case of amphibians tetrapod. In case of amphibians lungs are important respiratory organs. The gills are used for respiration only during larval stage. Therefore, aortic arches also show modifications as compare to fishes.

The urodel amphibians are aquatic inhabitant hence for respiration external gills are permanently present in them in addition to lungs. Therefore, their aortic arches show partial modification from condition in fishes. There are four pairs of arches (III, IV, V, and VI) usually present in urodels, although in some forms (*Siren, Necturns*) V arch is much reduced or absent. These amphibians show transition from 4 to 3 pairs of aortic arches.

The aortic arches are not broken by the external gills into afferent and efferent portions because branches arising from IV, V and VI aortic arches form capillaries in the external gills. The III arch forms the carotid arch. IV arch forms systemic arch. The lateral dorsal aortae between the III and IV aortic arches persist and each is known as ductus caroticus. From VI aortic arch a pulmonary artery grows out on either side taking blood to the lungs and skin. However, it also retains connection with radix or lateral dorsal aorta called *ductus arteriosus* or *ductus botalli*. In Anura and all amniotes the V aortic arch also disappears. Only III, IV, and VI aortic arches are present. The III arch along with a part of ventral aorta becomes the carotid arch. The IV arch along with its lateral dorsal aorta forms the systemic arch. The lateral dorsal aorta between the III and IV aortic arches (ductus caroticus) also disappears. The VI becomes pulmocutaneous arch and ductus arteriosus disappears during metamorphosis.

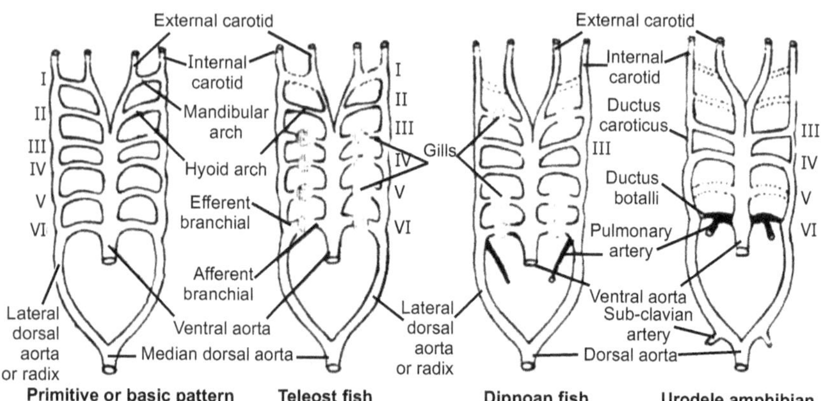

Fig. 5.7: Modifications of aortic arches in representative vertebrates

The larval anurans or tadpoles show arrangement of aortic arches similar to an adult urodeles. This is due to gill respiration. At metamorphosis, with loss of gills aortic arches I, II and V disappear altogether. Ductus caroticus also disappears so that the III or carotid arch takes oxygenated blood only to head region. IV or systemic arch on each side continues to dorsal aorta to distribute blood to elsewhere except head and lungs. Ductus arteriosus also disappears so that VI or pulmocutaneous arch supplies venous blood exclusively to lungs and skin for oxygenation. Thus adult anurans exhibit only three functional arches. i.e. III, IV and VI. which are also retained by higher vertebrates.

Reptiles: Reptiles are fully terrestrial animals in which gill or aquatic respiration is totally absent. The gills are replaced by lungs. Reptiles show only three functional arches i.e. III, IV and VI. But elongation of, shifting or heart towards posterior side and partial division of ventricle brings about certain modifications in the aortic system.

(i)   The ventral aorta and conus split into three trunks, two aortic or systemic trunk and one pulmonary trunk.

(ii) The right systemic arch (IV) arises from the left side of the ventricle while the left systemic carrying oxygenated blood to the carotid arch (III) to be sent into the head.

(iii) Left systemic arch (IV) take its origin from right ventricle carrying deoxygenated or mixed blood to the body through dorsal aorta.

(iv) Pulmonary trunk (VI) originated from right ventricle carries deoxygenated blood to the lungs for oxygenation.

(v) Ductus caroticus and ductus arteriosus are absent but in certain snakes and lizards (uromastix) ductus caroticus is present. In some turtles and sphenedon ductus arteriosus is present.

Birds: Birds are warm blooded animals because ventricles are completely divided into two compartments and there is no mixing of oxygenated and deoxygenated bloods. In birds the III, IV and VI aotic arches are present. They also follow the general pattern of reptiles with some differences. The ventricle shows complete division the conus arteriosus and ventral aorta have split to form two vessels, the systemic and pulmonary. Systemic trunk arising from the left ventricle and pulmonary trunk from right ventricle. The III form systemic trunk on the right side only. The part of the IV aortic arch of left side forms the left subclavian artery, the rest along with its lateral dorsal aorta disappears. The VI aortic arch forms pulmonary trunk.

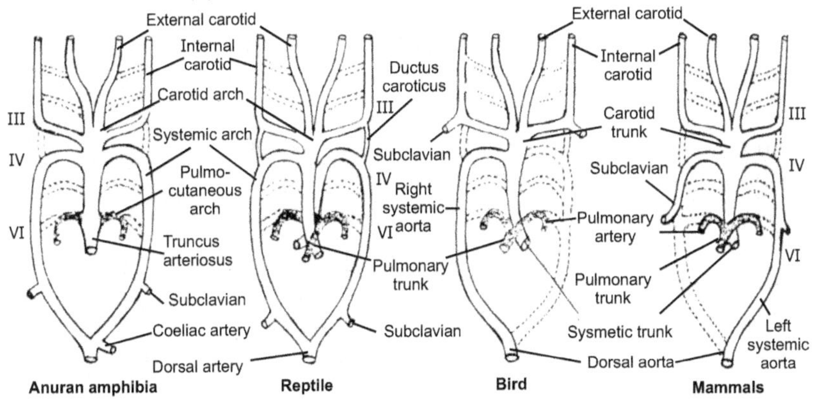

Fig. 5.8: Modifications of aortic arches in representative vertebrates

Mammals: In mammals also III, IV and VI aortic arches persist. The ventricle of mammalian heart is also completely divided into two chambers hence there is no mixing of oxygenated and deoxygenated blood. The conus arteriosus and ventral aorta split to form two aortae or trunks called systemic and pulmonary.

(i) The III aortic arch forms the carotid arteries.

(ii) The IV aortic arch forms the systemic trunk on the left side only emerging from the left ventricle and carrying oxygenated blood. While on the right side its proximal portion forms an innominate and right subclavian artery, the rest along with its lateral dorsal aorta disappears.

(iii) The VI aortic arch forms the pulmonary trunk taking deoxygenated blood from right ventricle to the lungs.

(iv) Embryonic ductus caroticus and ductus arteriosus also disappear. The latter persists in a reduced form as a thin ligamentum arteriosum.

Points to Remember

- Heart is important pumping organ.
- Heart is developed from mesenchyme.
- Single chambered, two chambered, three-chambered and four-chambered hearts are present in different vertebrates.
- In primitive vertebrates like amphioxus there are 60 pairs of aortic arches.
- In fishes there are seven pairs of aortic arches.
- In case of urodels four pairs of arches are present.
- Reptiles show three functional arches.
- In birds three arches are present.
- In mammals also three aortic arches are present.

Exercise

1. Give an account of evolution of heart in vertebrates.
2. Describe the different hearts in vertebrates.
3. How evolution of aortic arches occurs in vertebrates.
4. Give an account of evolution of aortic arches in vertebrates.
5. Write short notes on:
    (a) Development of heart
    (b) Single chambered heart
    (c) Two chambered heart
    (d) Three chambered heart
    (e) Four chambered heart
    (f) Aortic arches in amphibians
    (g) Aortic arches in reptiles, birds and mammals.

✱✱✱

Chapter **6**...

Urinogenital System

Contents ...

6.1 Succession of Kidneys in Vertebrates

Excretory or urinary system consists of kidneys and their ducts whereas reproductive system includes male and female gonads i.e. testes and ovaries and their ducts. The excretory organs kidneys excrete harmful metabolic wastes and regulate the composition of body fluids. Gonads produce the sex cells, eggs and sperms, and by the union of them zygote is formed. Thus, reproductive organs peretuate the species. Thus the organs of both the systems are functionally unrelated. But both the systems are closely related morphologically in vertebrates because male urinary ducts are used for both the purposes i.e. for carrying excretory waste and gametes. Therefore, both the systems together called *urinogenital* or *urogenital* system.

Vertebrate Kidneys and Ducts:

(1) Structure and Origin: There is a pair of compact kidneys lying dorsal to the coelom in trunk region one on either side of the dorsal aorta. Each kidney is made up of a large number of units called **uriniferous tubules** or **nephrons**. There number complexity and arrangement differ in different groups of vertebrates. The uriniferous tubules arise in an embryo from a special part of the mesoderm called **mesomere** or **nephnostome**. Primarily uriniferous tubules develop from the nephnostome in a sequence connecting from the anterior end, they are segmental in arrangement with one pair of uriniferous tubules from each trunk segment.

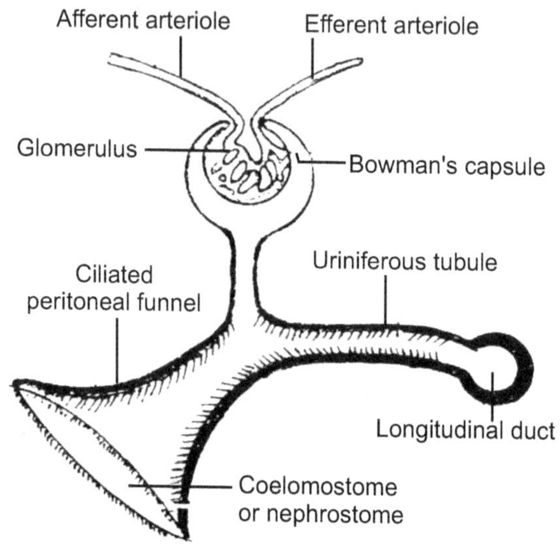

Fig. 6.1: Structure of embryonic kidney tubule

A uriniferous tubule is differentiated into three parts : peritoneal funnel, tubule and malpighian body.

(a) Peritoneal funnel: It is funnel-like ciliated structure present at the free end of the uriniferous tubule. It opens into coelum (splanchnocoel) by a wide opening called *nephrostome* or *coelomostome* for draining excretory products from coelomic fluid. Nephrostomes are usually confined to embryos and larvae and considered as vestiges of a hypothetical primitive kidney.

(b) Malpighian body: It is blind, cup like, hollow double walled Bowman's capsule. It encloses a tuft of blood capillaries called *glomerulus.* It get supply of blood from branch of renal artery called *afferent glomerular arteriole* and the out going branch is called *efferent glomerular arteriole.* It joins the capillary network surrounding the tubule.

Bowman's capsule and enclosed glomerulus together form a *renal corpuscle* or *malpighian body.* The glomeruli which are encapsulated are called internal glomeruli which are found in embryos, larvae and some fishes.

(c) Tubule: Malpighian bodies, water salts and excretory products from blood are carried by convoluted tubule where some important substances are reabsorbed. They open in embryonic cloaca through longitudinal duct. The segmented arrangement is lost and tubules become enclosed in a connective tissue capsule to form a kidney.

(2) Archinephros: The ancestral vertebrate had a pair of kidneys running through the entire length of the coelom, each had segmentally arranged tubule, one pair per body segment. Each tubule opens separately in the coelom by peritone funnel and nephrostome. Near each funnel, a glomerulus enclosed in a Bowman's capsule. The tubules of each kidney opens into a duct which joines the cloaca. This kidney is called complete kidney or *holonephros* and its duct is an *archinephric duct.*

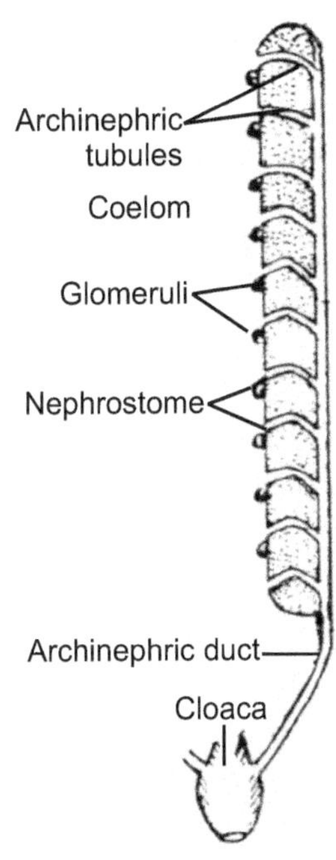

Fig. 6.2: Archinephros

Among the living vertebrates an archinephros is found today in the larvae of certain cyclostomes (*Myxine*) but not in any adult vertebrate. It is supposed to have given rise all the kidneys of later vertebrates during the course of evolution. Modern or present day vertebrates show three types of adult kidneys namely, *pronephros, mesonephros,* and *metanephros.* These three types are nothing but the three successive stages of development from the ancestral kidney i.e. *achinephros* as all these three types were not functional at the same time.

Pronephros:

In the embryos of all vertebrates, the first kidney tubules appear dorsal to the anterior end of coelom, on either side. These kidneys are called pronephros because they are first to appear. These are also called as head kidney because of its anterior position immediately behind the head. Pronephros develops in the anterior most part of the nephrostome. It consists of 3 to 15 uriniferous tubules segmentally arranged one pair to each segment. There are only three pronephric tubules in frog embryo, seven in human embryo and about a dozen in chick embryo. Each tubule opens into coelom by a funnel or nephrostome. Near each tubule is a glomerulus without Bowman's capsule and peritoneal funnel, called naked or external glomerulus. In some cases of pronephros, the glomeruli are surrounded by Bowman's capsule and they are called internal glomeruli. In some cases, glomeruli unite to form a single compound glomerulus called *glomus*. Glomus and tubules become surrounded by a large *pronephric chamber* derived from pericardial or pleuroperitoneal cavity. Originally each tubule has its individual external aperture, but secondarily, all tubules of a pronephros open into a common pronephric duct leading posteriorly into the embryonic cloaca.

In all vertebrate embryos and larval stages, the pronephros is functional and it is mostly transitory and soon replaced by the next stage or mesonephros. However, a pronephros is retained throughout life in adult cyclostomes and few teleost fishes, but it is non-urinary and mostly lymphoidal in function. The pronephros which become adult kidney is some vertebrates called head kidneys. Head kidney has a mass of uriniferous tubules opening by nephrostomes into the pericardial cavity and it has one glomus. e.g. **Myxine** and some teleosits.

Mesonephros:

Mesonephros develops from the part of the nephrostome which lies behind the pronephros. In embryos, it develops from middle part of intermediate mesoderm posterior to each pronephros soon after its degenaration.

At first it consists of paired segmental uriniferous tubules, each with peritoneal funnel opening into the coelom and a glomerulus enclosed in a Bowman's capsule. The mesonephric uriniferous tubules join the exisiting pronephric ducts on each side which is now called *mesonephric duct* or *Wolffian duct*. Later on the mesonephric tubules undergo budding to form hundreds of tubules so that their segmental arrangement is lost. The later mesonephric tubules may have no pentanel funnels. Mesonephros form the adult functional kidneys in some cyclostomes, in fishes and amphibians. They also form the kidneys of embryos of amniotes in which they degenerate in the adult. The tubules of pronephros and mesonephros develop similarly and are homologous. However, mesonephros is functionally better and advanced than pronephros because pronephric tubules are in large number, longer and develop internal glomeruli enclosed in capsules forming malpighian bodies, They remove liquid wastes directly from glomerular blood instead of coelomic fluid as in the case of a pronephros. Mesonephros is also called *Wolffian body*. As the pronephros disappears, the old pronephric duct becomes mesonephric duct or wolffian duct.

Mesonephros is functional in embryos of reptiles, birds and mammals. But in adults they are replaced by metanephros. In case of fishes and amphibians mesonephros is functional both in embryos and adults. In shorns and caecilians, tubules extend posteriorly throughout the length of coelom. Such kidney is called *posterior kidney* or *opisthonephros*. Whereas in adult anurans, urodeles and embryonic amniotes, the mesonephros does not extend posteriorly. Mesonephric kidney is not metameric but in myxinoids it is segmental and some times called a holonephros.

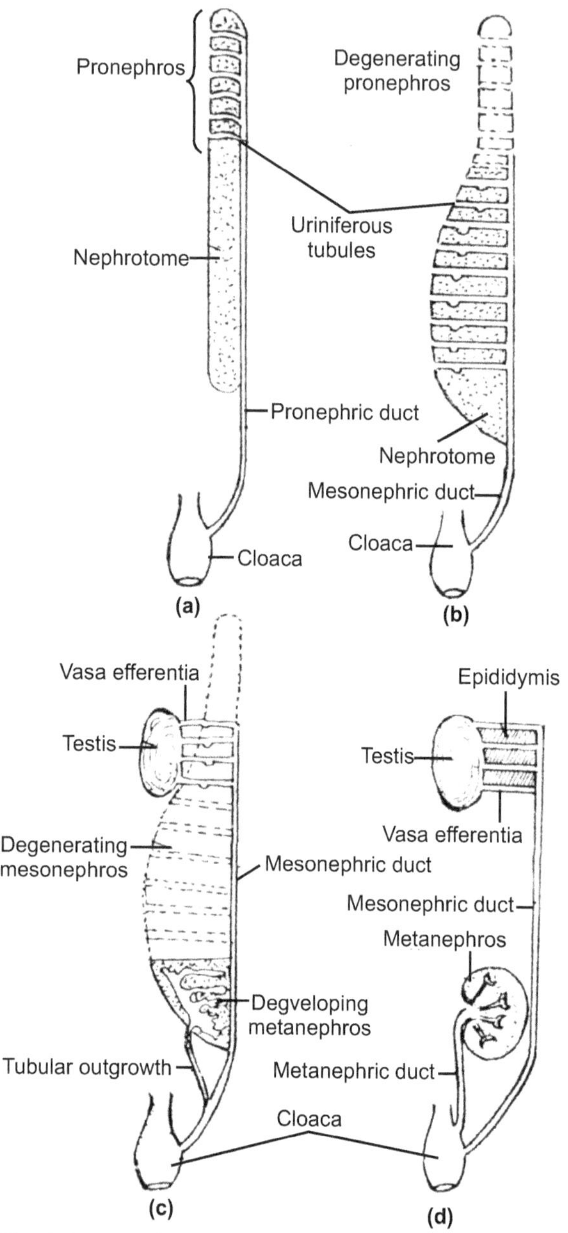

**Fig. 6.3: Development of kidneys and their ducts
(a) Pronephros, (b) Mesonephros, (c) Metanephros forming and mesonephros joining testis, (d) Metanephros**

Metanephros:

It is the functional kidney of higher vertebrates or amniotes. which is formed from the posterior part of the nephrostome behind the embryonic mesonephros. Metanephric kidney has a double origin, a tubular out growth arises from the mesonephric duct near the cloaca and it grow into the nephrostome where it divides into branches, the branches form collecting tubules and calyces. The proximal part of the tubular outgrowth becomes the *metanephric duct* or *ureter*. The nephrostome gives rise to metanephric thousands of uriniferous tubules but they do not show segmental arrangement. The metanephric tubules become long and much coiled. They are with glomeruli enclosed in Bowman's capsules. They lack peritoneal funnels so that all connection with the coelom is lost. The metanephric kidneys are found in higher vertebrates and they have achieved the separation of the urinary function from the genital function which appears to be the trend in the evolution of urinogenital system. There are three processes in the excretion, namely, *filtration* of blood in the glomerus, *secretion* of certain waste substances by the cells of uriniferous tubules into the lumen of the tubules, and *selective reabsorption* by uriniferous tubule.

In case of aquatic vertebrates, kidneys chiefly eliminate excess of water absorbed in the body, they excrete ammonia which is diluted with water and made non-toxic. In case of terrestrial vertebrates, kidneys conserve water for maintaining water balance of the body and they excrete urea and uric acid.

In case of mammals metanephric kidney show highest degree of organisation with several feathers. There is a thin U-shaped loop of *Henle* present between proximal and distal convolutions of a metanephric tubule. Such loops are absent in reptiles and rudimentary in birds. Kidney shows outer cortex containing renal corpuscles and an inner *medulla* having collecting tubules and loops of Henle; which are aggregated into one or several pyramids tapering into pelvis.

Urinary bladder:

In most vertebrates, bag like urinary bladder is present to store urine before it is discharged. However, it is lacking in cyclostomes, elasmobrachs, some lizards, snakes, crocodiles and most birds. In case of fishes, bladder is formed by an enlargement of the mesonephric ducts and it is called as tubal bladder. In Dipnoi, it evaginated from dorsal wall of cloaca. In tetrapods, it evaginates from the ventral wall of cloaca. It is called cloacal bladder in amphibian. In amniotes, the adult bladder is derived from the proximal part of embryonic allantois and hence it is called allantois bladder. Generally, the kidney ducts or *ureters* do not open into the urinary bladder except in mammals. They open into cloaca.

6.2 Gonads and their Ducts

There is sexual reproduction in vertebrates and sexes are separate with few exception of hagfishes and few bony fishes. They have hermaphrodite gonads. Testes are male reproductive gland which produce sperms or male gametes. Ovaries are female gonads and they produce ova or eggs. In embryo, gonads originate as a pair of thick elevated folds or genital ridges of coelomic epithelium from the roof of coelom, one on either side of the dorsal mesentry. Genital ridges are much longer than the functional adult gonads, suggesting that in the ancestral vertebrates the gonads extended the whole length of the pleuroperitoneal cavity. The functional adult gonad is derived from the middle or gonal part of the genital ridge, while its anterior progonal and posterior epigonal parts remain sterile. Gonads remain suspended in coelom from dorsal body wall of a fold of dorsal mesentery, called *mesorchium* in male and *mesovarium* in females. Generally one pair of gonads is present. But some vertebrates have a single gonad only because of either fusion of both embryonic genital ridges (most cyclostomes, perch and some other fishes) or degeneration of one juvenile gonad (hag fishes, some elasmobranchs and lizards, alligators and most birds). Associated with the gonads are special gonoducts or genital ducts, *vasa deferentia* in males and oviducts in females, to transport gametes to cloaca or to outside body. However, cyclostomes and few elasmobranchs lack genital ducts. Their sperms and eggs escape body cavity via. abdominal pores.

1. Testes and Male Genital Ducts:

Testes of vertebrates are paired organs of moderate size usually found attached to kidneys. Each testis is a compact gland, covered by coelomic epithelium and composed of numerous highly coiled *seminiferous tubules* embedded in the connective tissue. Tubules are lined by germinal epithelium which gives rise to billions of sperms. On maturity sperms are set free in the lumen of tubules and move towards the genital ducts.

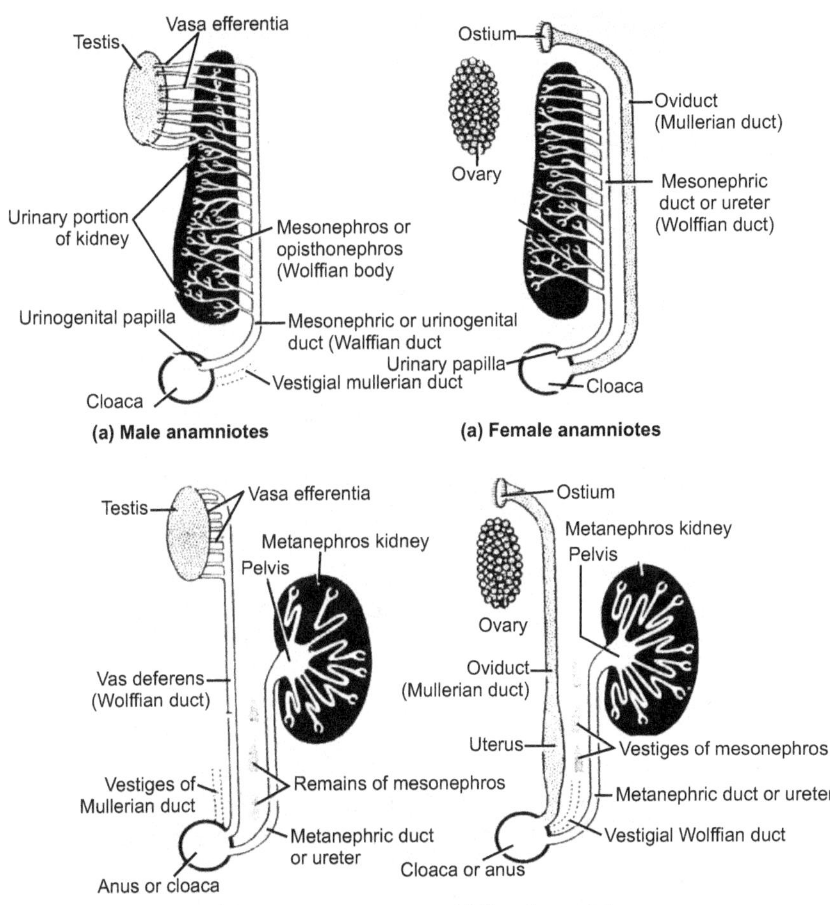

Fig. 6.4: Schematic representation of evolution of urinogenital organs and their ducts in vertebrates

Some cyclostomes have a single median testis without genital duct. Sperms are released in the coelom, from where they pass through abdominal pore, located at posterior part of coelom. In dog fish, two testes are elongated bodies. In most amniotes, the opisthonephros (or mesenephros) is differentiated into anterior genital portion sem in males, some uriniferous tubules lose excretory function; from slender vasa efferentia and become continuous with seminiferous tubules of the adjacent testis. They serve to convey sperms of testis to the mesonehric duct of kidney. Thus, in male amniotes mesonephric or Wolffian duct forms a urinogenital duct, serving both as a vas deferens for sperm as well as a ureter for urine.

However, in many elasmobranchs (e.g. dogfish) accessory urinary ducts drain urine from kidney to cloaca so that the mesonephric duct serves entirely as a vas deferens. The anterior genital part of kidney along with the part of mesonephric duct from an *epididymis.*

In Anuran embryo, each testis is made of two portions. In male frog anterior portion disappears and posterior portion becomes the adult functional testis.

In adult male toad, the anterior portion also persists as Bidder's organ, containing large cells similar to immature ova.

In male amniotes, a metanephros develops as the adult functional kidney with its own urinary duct or ureter to transport urine. Thus, mesonephric or Wolffian duct becomes solely a *genital duct* or *vas deferens.* The remnants of embryonic mesonephros and coiled portion of mesonephric duct becomes *epididymis* of the adult kidney. From each testis sperms pass first through epididymis, then through vas deferens to reach urethra.

In most mammals testes descend permanently into extra abdominal skin bags called scrotal sacs. In rabbits, bats and rodents, they are lowered into sacs and retracted at will. Passage between abdominal cavity and scrotal sac, through which testes descends, is called inguinal canal. However, some mammals such as monotremes, insectivores, elephants, whales etc. lack scrotal sac so that their testes remain permanently intra-abdominal like ovaries.

2. Copulatory organs:

The amniotes who have external fertilization do not have copulatory organs. But these show fertilization have the copulatory organ for mating or copulation. They are prominent in reptiles and mammals. In elasmobranchs (e.g. dogfish) bases of pelvic fin are modified as intromittant organs called claspers. These are grooved, cylindrical structures that are inserted into the female cloaca to inject sperm. In several teleosts, the anal fin is modified as a *gonopodium* for sperm transport. Snakes and lizards have a pair of retractile, grooved and sac-like hemispenes which can be everted through cloaca. The retraction is controlled by modified body wall musculature.

Turtles, crocodilians, some birds and prototherian mammals have unpaired grooved and erectile penis formed as a thickening of cloacal floor. Only higher mammals have a true external, erectile penis with a tubular groove continuous with a spongy urethra. A series of accessory sex glands associated with penis secrete a fluid in which sperms are carried.

3. Ovaries and Female Genital Ducts:

In female anamniotes, ovaries are large, occupying much of the body cavity and produce thousands of eggs as fertilization is external. In amniotes ovaries produce fewer eggs because fertilization is internal. Ovaries of reptiles and birds are still large and the eggs produced contain much yolk. However, mammalian eggs contain very like yolk so that their ovaries also remain quite small.

Ovaries are generally paired structures, but only a single median ovary occurs in cyclostomes, as also in some teleosts (e.g. perch). They are not attached to the kidneys like testes in the males. Only the right ovary in functional in many elasmobranchs whereas only the left ovary becomes mature in birds and primitive mammals.

In all vertebrate embryo, except cyclostomes, the coelomic epithelium on the outside of mesonephric duct develops groove which becomes closed to form a tubule called *Mullerian duct*. In adult males, mullerian duct becomes vestigial and functionless. In adult females, the Mullerian duct grows larger and becomes the female genital duct or oviduct. It opens anteriorly into coelom, in the region of degenerating pronephros, by a *coelomic funnel* or *ostium* and terminates posteriorly into cloaca. In female elasmobranchs, the Mulleiran duct is formed differently by the longitudinal splitting of the pronephric duct. Thus, in adult female anamniotes, both Mullerian duct (oviduct) and the Wolffian duct (mesonephric or urinary duct) are present. But, in adult female amniotes, with the development of adult metanephros and its metanephric duct or ureter, the mesonephros and its duct (Wolffian duct) degenerate leaving only vestiges known as *provarium*.

In viviparous mammals, posterior ends of both the Mullerian ducts become fused and are modified into a uterus in which the embryos develop and a *vagina* which receives the male intromittant organ during copulation. The remaining anterior parts or oviducts are relatively short, narrow and convoluted and called the fallopian tubes. Condition of uteri varies in different mammals. When uteri remain double without fusion, it is called *duplex uterus* (marsupials). When uteri partially fuse so as to form two horns and two separate lumens inside, it is called *bipartite* (hamster, rabbit). When there are two horns but a single internal cavity it is called *bicarnnate uterus* (ungulates). When uterine horns are absent and both uteri fuse completely with a single internal cavity, it is termed *simplex uterus* (Primates, some bats and armadillos).

Points to Remember

- The system which works for both excretory and reproductive function called urinogenital system.
- Kidneys are excretory organ and contain large number of uriniferous tubules or nephrons.
- Uriniferous tubule show three parts, peritoneal funnel, tubule and malpighian body.
- Complete kidney has archinephric duct.
- The kidneys are called pronephros as they are first to appear.
- Mesonephros kidneys are formed in cyclostomes, fishes and amphibians.
- Metanephros is the functional kidney of higher vertebrates.
- In most vertebrates bag-like urinary bladder is present to store urine before it is discharged.
- Sexes are separate in higher vertebrates.
- Testes produce sperms and ovary produce ova or eggs.
- General gonads are present in pairs.
- Some vertebrates have single gonad (e.g. hagfish, sharks, lizards, alligators and birds).
- Vas deferens or Wolffian duct carries sperms from testes.
- Oviduct or Mullerian duct carries ova from ovary.
- In many vertebrates copulatory organ is present useful for internal fertilization.

Exercise

1. Give general account of evolution of kidney in vertebrate animals.
2. Trace the fate of pro-, meso-, and metanephros in vertebrates.
3. Describe the evolution of genital ducts in different vertebrates.
4. Write short notes on:
   - (a) Archinephros
   - (b) Mesonephros
   - (c) Metanephros
   - (d) Mullerian duct
   - (e) Structure of embryonic kidney tubule
   - (f) Urinary bladder
   - (g) Testes and male genital ducts
   - (h) Copulatory organs
   - (i) Ovaries and female genital ducts.
5. Give an account of testes and male genital ducts.
6. Give an account of ovaries and female genital ducts.

Chapter **7**...

Nervous System

Contents ...

7.1 Brain of *Scoliodon*

Brain: The brain is divided into three parts fore-brain, mid-brain and hind-brain.

Fore-brain: The fore-brain includes olfactory lobes, cerebrum and the diencephalon.

The olfactory lobe consists of pair of stout stalks, the olfactory penduncles extending forwards and outwards from the anterolateral angle of the cerebrum. Olfactory peduncles end in a bilobed mass called as olfactory bulb or lobe, closley applied to olfactory sac of its own side. The olfactory tract and bulbs enclose narrow cavities, the olfactory ventricles or rhinocoels.

The cerebrum is undivided massive structure with no median groove to separate it into right and left cerebral hemispheres. It contains pair of cavities called *lateral ventricles* or *paracoels* separated by median partition and they are continued with *foramen* of *Monro*. The dorsal surface of cerebrum is quite smooth but on mid ventral surface, there is small opening, called neuropore through which emerges a pair of delicate nerves, the terminal or pre-olfactory nerves. Each of these nerves bears a ganglion along its course and runs alongside the olfactory tract of its own side to innervate the mucous membrane of the olfactory sac.

The cerebrum is continued behind into the narrow diencephalon which is very short and is completely hidden by the forward prolongation of the cerebellum. The roof of the diencephalon is extremely thin and membranous, being non-nervous in character but it contains numerous pineal stalk arises from the roof and runs forward and upward to the membrane, covering the anterior fontanelle of the skull. The diencephalon encloses laterally compressed cavity, the third ventricle or diacoel that is continuous behind with the iter. The floor of the diencephalon gives off a hollow outgrowth, the infundibulum which projects downward and backward. Close to the lateral sides of infundibulum, lie two thick walled oval sacs, the lobi inferiores, the distal ends of which are continued into a pair of glandular sacs, the *sacci vasculosi* with thin walls. The median outgrowth is derived from the roof of the buccal cavity, called *hypophysis*, which joins the distal end of the infundibulum. In front of the infundibulum, lies the optic chiasma, the crossing of the *optic nerves*.

Mid-brain: The mid-brain consists of optic lobes. These are two large oval bodies on the dorsal side of the brain. They are covered dorsally by the cerebellum and ventrally by infundibular outgrowths. Each optic lobe has a cavity called *optocoel* which opens into the iter.

Hind-brain: The hind-brain comprises of cerebellum and medulla oblongata.

The cerebellum is very large, elongated antero-posteriorly parts of the optic lobes and medulla. Its dorsal surface is thrown into numerous irregular folds. Two deep transverse furrows divide the cerebellum into three lobes. It has the cavity called metacoel, that opens below into iter. From its anterior end arises, a pair of hollow outgrowths, the *carpora restiformia* or *auricular lobes*. The medulla oblongata is triangular, with thin roof forming posterior choroid plexus. It encloses the wide fourth ventricle or *myelocoel*. The medulla oblongata is continuous with spinal cords.

All the ventricles of the brain are continuous and filled with fluid called *cerebro-spinal fluid*, which is secreted by anterior and posterior *choroid plexus*.

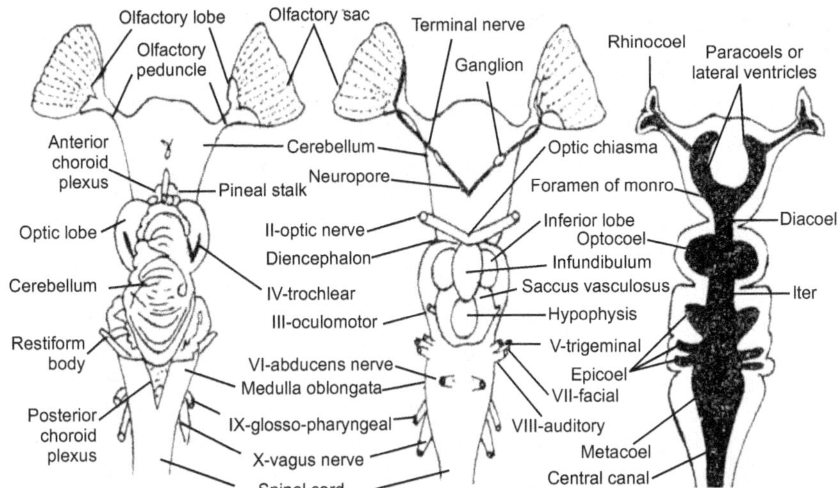

(a) Dorsal view, (b) Ventral view, (c) Horizontal longitudinal section
Fig. 7.1: *Scoliodon*. Structure of brain

Functions: Olfactory lobes and cerebrum are useful for sense of smell. The cerebellum is the seat of regulation of balance and muscular control. The hypothalamus is concerned with gustatory and *olfactory* impulses and with the control of visceral functions. The lobi inferiores and sacci vasculosi are centres for smell and taste. The optic lobe has optic, olfactory, gustatory and acoustico-lateral sensory centres. The medulla oblongata contains the respiratory centres besides the centres of number of involuntary activities.

7.2 Brain of Frog

The brain and the spinal cord together form the central nervous system. The system is hollow with tubular structure situated along the mid-dorsal axis of the animal.

(A) Brain:

Brain is the main controlling and co-ordinating centre of all body activities. It is the anterior, well developed and well differentiated part of the central nervous system. It is protected in a hard bony box called *cranium.* Inside the cranium brain is further enclosed by three membranes called *meninges.* The outer brown fibrous tough and protective membrane which lines the cranium called *dura matter.* The innermost membrane closely applied to the brain is the thin, delicate, and pigmented called *pia matter.* The narrow space between the membranes and the inner cavities of brain are filled with a clear, watery *cerebrospinal fluid.*

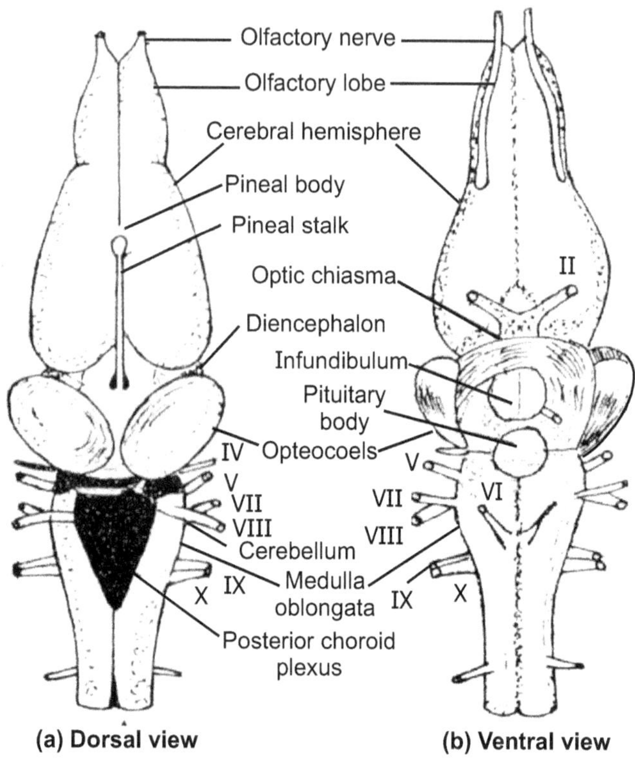

Fig. 7.2: Brain of Frog

The meninges protect the brain from mechanical shocks and injuries. The cerebrospinal fluid works as a lubricant and shock absorber. This cerebrospinal fluid supplies oxygen and is also nutritive function. It prevents desiccation of brain and also helps feebly rigid brain to withstand the stress. The nervous tissue of the brain is differentiated into inner grey matter which is surrounded by white matter.

The brain of the frog is differentiated into three main parts, namely:

(1) Fore-brain or Prosencephalon
(2) Mid-brain or Mesencephalon and
(3) Hind-brain or Rhombencephalon.

(1) The Fore Brain or Prosencephalon:

It is the anterior part of the brain and consists of olfactory lobes, cerebral hemispheres and diencephalon or thalamencephalon.

(a) Olfactory Lobes: These are a pair of small lobes projecting anteriorly from the cerebral hemispheres. Dorsally they are completely fused with each other but differentiated from each other by longitudinal and transverse grooves. Each olfactory lobe sends olfactory nerve to the nasal chamber of its side. They are with sensory nerves. Internally each lobe contains a small cavity, the olfactory ventricle or rhinocoel.

The olfactory lobes are concerned with the sense of smell.

(b) The Cerebral Hemispheres: The cerebral hemispheres or cerebrum lie behind the olfactory lobes. They are separated by a shallow transverse groove. They forms the major parts of the brain. These are long, oval and smooth structures, narrow infront but broad behind and separated from one another by a deep mid-longitudinal groove or fissure. Each hemisphere encloses a cavity called the lateral ventricle (or the left first and the right second ventricle). They are also called as paracoels and are continuous anteriorly with the olfactory ventricles. Posteriorly, they unite with each other and with diocoel of diencephalon by a common passage, the foramen of Monro. The floor of ventrolateral walls of the lateral ventricles are thickened to form the *corpora striata (singular corpus striatum)*. The two corpora striata are interconnmted by a tract of transverse fibres called the *anterior commissure*. The roof and dorso-lateral walls of the lateral ventricles are relatively thin and known as the *pallium*.

The cerebral hemispheres function as the seat of memory, intelligence and decision. It also controls voluntary actions (will power) of the body. The corpus striatum regulates the body temperature in cold blooded animals.

(c) Diencephalon: Behind the cerebral hemispheres lies an unpaired diencephalon. It is the short, rhomboid, depressed region on the dorsal surface of the brain. The dorsal and the ventral walls of the diencephalon are thin, but the lateral walls are thick and bulging called optic thalami (singular thalamus). The roof of the diencephalon lies anterior to choroid plexus. It is non-nervous and highly vascular structure. From the posteriodorsal side of the diencephalon arises a hollow stalk known as *pineal stalk*. It bears at its free end a knob like structure called *pineal body*. It is considered to be a vestigal third eye.

On the ventral side of the diencephalon is a median bilobed projection called *infundibulum*. Attached to the ventral surface of the infundibulum is a small rounded structure called hypophysis or Rathke's pouch. The infundibulum and hypophysis collectively form the pituitary body. Anterior to the pituitary body optic nerves cross each other forming X-shaped structure called *optic chiasma*. Diencephalon encloses the third ventricle of diocoel. The floor of the third ventricle is called the *hypothalamus*.

The function of diencephalon is to control the spontaneous (involuntary) movements of the body. It also regulates body temperature reproductive activities and sleep. The pituitary gland is the most important endocrine gland controlling all other endocrine glands. It secretes number of hormones. The secretion of pituitary body control the general growth of the body as well as pigmentation skin. Diencephalon also works as a relay centre for conducting impulses to cerebral hemispheres. The pineal body which is functionless vestigeal third eye. The optic thalami of diencephalon act as association centres for the sense of sight.

(2) Mid-brain or Mesencephalon:

It forms the middle part of the brain and consists of the optic lobes and crura cerebri.

(a) Optic lobes: The optic lobes are also known as the *corpora bigemina* (singular - corpus bigeminum). They are pair of prominent oval and hollow bodies diverging anteriorly and converging posteriorly situated on the dorsal side of the brain. They, however, project on the sides so that they are partly visible from the ventral side also. Each optic lobe contains a cavity called the optocoel. Both the optocoels are connected with each other and also to the third ventricle in front and the fourth ventricle behind by narrow space called iter or *aqueduct* of *Sylvius*. From the ventral side of each optic lobe arises the optic nerve. The optic nerves of either side cross each other in front of infundibulum to form the optic chiasma.

The optic lobes are concerned with the sense of sight, co-ordination of movements and checking the spinal reflexes. Because of crossing of optic nerves, each optic lobe controls the opposite side of the body.

(b) Crura cerebri: There are pairs of thick longitudinal bands of nervous tissue situated ventrally on the floor of the iter structurally and functionally connect forebrain to hindbrain transmitting impulses to and fro.

(3) Hind brain or Rhombencephalon:

It consists of cerebellum and medulla oblongata.

(a) Cerebellum: The cerebellum is poorly developed narrow transverse solid ridge or band, placed dorsally just behind the optic lobes.

The function of cerebellum is to maintain the equilibriurn of the body and along with cerebrum it controls and co-ordinates muscular movements of the body.

(b) The Medulla Oblongata: It is the posterior most part of the brain. It is slightly broader at the anterior end and narrow posteriorly where it is continued into spinal cord without any line of demarcation. On the dorsal side it shows a highly vascular patch of pia matter called posterior choroid plexus. It nourishes the posterior part of the brain. The medulla oblongata encloses fourth ventricle or metacoel which is continued behind in the spinal cord as the cerebrospinal canal. The floor of the fourth ventricle is very thick.

As regards the functions, the medulla oblongata controls most of the important involuntary vital functions of the body such as, heart beats, peristalsis, blood pressure, respiratory movements etc. It is also concerned with catching and swallowing of prey, croaking, digestive reflexes like secretion of gastric and pancreatic juices.

The posterior choroid plexus secretes cerebrospinal fluid and provides nourishment as well as oxygen to the internal parts of hind brain by giving offshoots. The medulla oblongata forms a bridge connecting the spinal cord in the brain. The removal of medulla oblongata or damage to it soon results in death.

Ventricles of Brain:

The brain is hollow, its cavity being enlarged or reduced at places to form ventricles. They are filled with cerebrospinal fluid. Frog brain shows the following ventricles.

(1) Rhinocoels or Olfactory Ventricels: These are pair of cavities in the olfactory lobes.

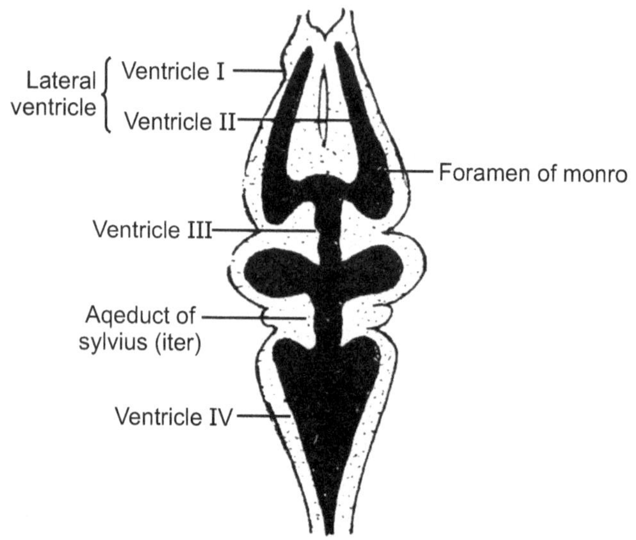

Fig. 7.3: Showing ventricles

(2) Paracoels or Lateral ventricle: These are pair of narrow ventricles present in the cerebral hemispheres. They open into third ventricle behind through a common aperture called foramen of Monro.

(3) Diecoel or Third ventricle: It is a narrow median ventricle present in diencepalon.

(4) Opteocoels or Optic ventricles: These are present in the optic lobes. They are connected to the third ventricle in front and fourth ventricle behind through a narrow space called 'Iter' or 'Aqueduct of Sylvius'.

(5) Iter or Aqueduct of Sylvius: It is a narrow space connecting the optocoels as well as the third ventricle infront and the fourth ventricle behind.

(6) Metacoel or Fourth ventricle: It is the spacious cavity present in the medulla oblongata. It posteriorly continues into the spinal cord as the cerebrospinal or central canal.

7.3 Brain of *Calotes*

The brain and the spinal cord constitute the central nervous system.

Brain: The brain is enclosed in the cranial cavity of the skull. The nervous tissue consists of the nerve cells called neurons. The neurons give fibrous elements and non-nervous supporting cells called *neuroglia*. All these structures are ectodermal in origin. The cellular elements form grey matter, while fibrous elements form the white matter in the nervous tissue. In the brain grey matter is dorsal and peripheral in position, while the white matter is confined mainly to the ventral area. The brain is covered and protected by connective tissue sheaths called meninges. The nervous tissue of the brain is protected by two meninges called pia matter and dura matter. The pia matter remains in close contact with brain and it is highly vascular. The dura matter lies just outside the pia matter and is mainly fibrous in nature. The two coverings remain separated from each other and the space between them is called subdural space.

The brain of the Calotes is differentiated into:

(a) Fore brain, (b) Mid brain and (c) Hind brain.

(a) Fore brain: It comprises of olfactory lobes, cerebral hemispheres and diencephalon.

(1) Olfactory lobes: The olfactory lobes form the anterior most part of the brain. Each lobe consists of a posterior long, narrow olfactory penducle or tract and an anterior slightly swollen olfactory bulb. From the olfactory lobes arise olfactory nerves which innervate the nasal capsule. The entire olfactory lobe contains a cavity called olfactory ventricle or rhinocoel. The olfactory lobes control the sense of smell.

(2) Cerebral hemisphere: These are pair of large, oval, smooth bodies broad behind and narrow in front. The two hemisphere lie in close contact with each other in the median line. A dorsal groove, the median longitudinal fissure however, forms the demarcation between the two lobes. The cerebral hemispheres are hollow inside and the cavities present in each hemisphere is called lateral ventricle (1st and 2nd ventricles) or paracoel. These cavities are in continuation with the rhinocoel cavities of olfactory lobes and filled with cerebrospinal fluid. The roof of the cerebral hemisphere is very thin called the pallium but the floor is very thick formed by 4-5 layers (strata) of grey and white matter.

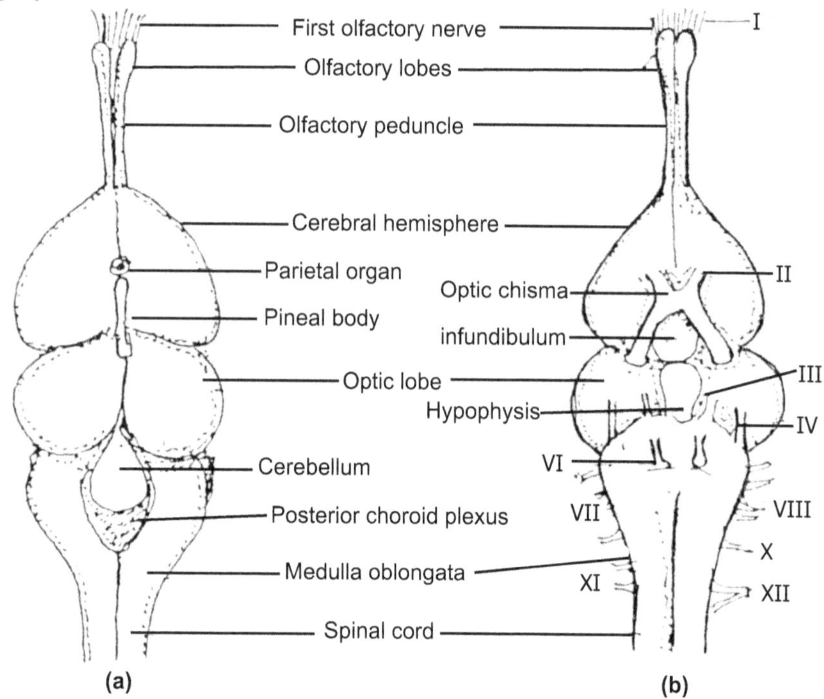

Fig. 7.4: *Calotes versicolor*
(a) Brain dorsal view and (b) Ventral view

The lateral and ventral walls of the cerebral hemisphere are thick because of these layers together called corpus striatum. On the ventral side longitudinal fissure demarcates the *hippocampal* lobes on either side. Hippocampus is the swollen part of the midposterior region of the cerebrum. Two sides of the brain are connected at places by transverse bands of nerves are called commissures. The cerebral hemispheres are connected by two commissures. Anterior commissure is the wide transverse tract of the nerve fibres connecting two corpora striata and receive many efferent fibres from more functions to the cerebrum. Just above the anterior commissure, hippocampal commissure connects the hippocampal regions of the cerebral hemispheres. The cerebral hemispheres are the seat of intelligence, memory, will and thought. They also control voluntary actions in higher animals like mammals. But in Calotes cerebrum retain olfactory role in these animals. The corpora striata play an important role of cerelation and co-ordination centres.

(3) Diencephalon: The diencephalon is a small, laterally compressed narrow area situated between cerebral hemisphere anteriorly and optic lobes posteriorly. Dorsally, it is almost invisible due to overlapping by the cerebral lobes and the optic lobes. It encloses a cavity, the third ventricle or diacole which communicates with the lateral ventricles of the cerebral hemisphere a small comman aperture, the foramen of Monro. The roof of the diencephalon is very thin and it is called the epithalamus. Anteriorly the roof is very thin and greatly folded to form a much branched and deeply vascular structure called the anterior choroid plexus. It secretes the cerebro spinal fluid. Behind the plexus is found the epiphyseal or pineal apparatus, which consists of two parts : the anterior eye like parietal body or pineal eye and posterior pineal body. The pineal eye is the functionless vestigial third eye. It was earlier a functional eye opening out through parietal foramen of the cranium. It is still able to perceive light in some lizards, e.g. Varanus, Anguis and lizard like Sphenodon. It registeres solar radiation and influences animals behaviour to expose itself to warm sun. The pineal body is elongated sac-like gland and it is endocrinal in nature and secretes melanin which causes the pigment granules to clump, thus turning the skin to pale.

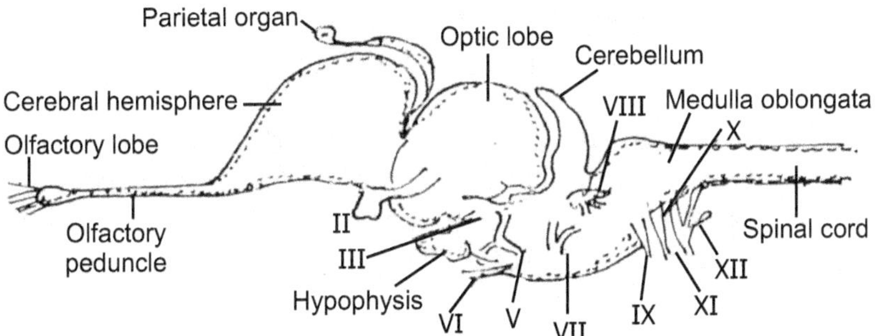

Fig. 7.5: *Calotes versicolor* - Brain lateral veiw

The posterior and Habenular commissures occur behind the pineal apparatus. The side walls of the third ventricle are thick and are called thalami or optic thalami in which many of the fibres of optic nerve end. Functionally, the thalami serve as relay centres for the transmission fibres between the corpora striata in front and the hind brain and spinal cord behind. The floor of the third ventricle is also thick and shallow called the hypothalamus. The hypothalamus exerts a partial control over the smooth muscles and glands of the body through the autonomic nervous system. Ventrally, the hypothalamus is produced into a funnel shaped structure called the infundibulum. To the distal end of infundibulum is attached the hypophysis. These two structures together constitute pituitary body or hypophysis. Pituitary gland which is the most important endocrine gland controlling all endocrine glands. It secretes number of hormones. In front of this body lies the crossing of the optic nerves the optic chiasma. The diencephalon contains centres for spontaneous movements.

(b) Mid brain: The mid brain or mesencephalon consists of optic lobes and crura cerebri.

(1) Optic lobes: The optic lobes are pair of large rounded bodies on the dorsal side in contact with cerebral hemispheres in front. These lobes arise as dorsolateral outgrowths from the floor of mid brain. From the ventral surface the optic lobes give rise to two optic nerves which cross each other to form optic chiasma in front of infundibulum. Each optic lobe encloses a cavity called optocoel or optic ventricle. Both optocoels open into a narrow median longitudinal canal called iter or aquaduct of Sylvius. It connects the third ventricle with the fourth. The optic lobes are concerned with the sense of sight.

(2) Crura cerebri: The crura cerebri are thick, longitudinal bands of nerve fibres below the optic lobes, connecting the diencephalon with the medulla oblongata. Through crura cerebri to and fro movements of impulses between the cerebrum and spinal cord.

(c) Hind brain: Hind brain or metencephalon consists of mainly two parts and they are cerebellum and medulla oblongata.

(1) Cerebellum: The cerebellum is poorly developed structure. This is perhaps due to the fact that these animals move mainly in one plane and show sluggish behaviour. It consists merely of a narrow, flat, transverse band on the dorsal side just behind the optic lobes. It lacks lateral lobes. Cerebellum is solid, undivided flap of grey matter which covers the optic lobes. Cerebellum controls and co-ordinates muscular movements and thus controls equilibrium of the body.

(2) Medulla obongata: It is also called *myelencephalon* and it is the posterior most part of the brain and continues behind as the spinal cord. It is broad in front and narrow behind. It has a prominent ventral flexure, where it passes into the spinal cord. The roof of this region is very thin, membranous and deeply vascular, It is known as the posterior choroid plexus. It secretes the cerebro-spinal fluid and supplies nourishment to the posterior part of the brain. Medulla oblongata encloses a cavity, the fourth ventricle or myelocoel. The sides and the floor of the medulla oblongata are very thick and serve as reflex centres for the auditory and visceral nerves. Cranial nerves from 5 to 12 take their origin from this region.

7.4 Brain of Pigeon

The brain of pigeon is relatively larger than the lower forms like fish, amphibians and reptiles and is comparable with that of mammals. But there are no superficial convultions which are found in mammals.

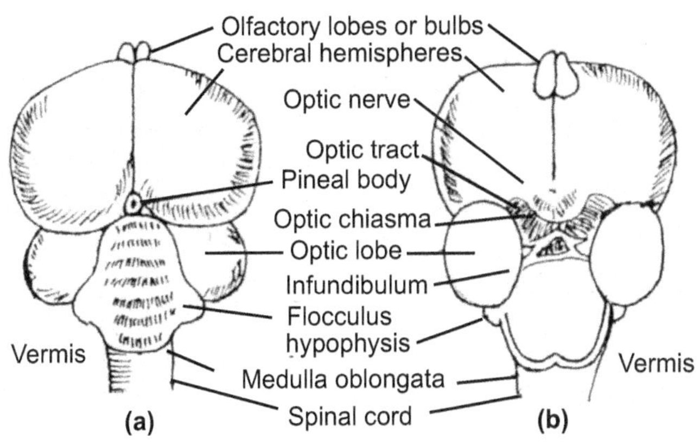

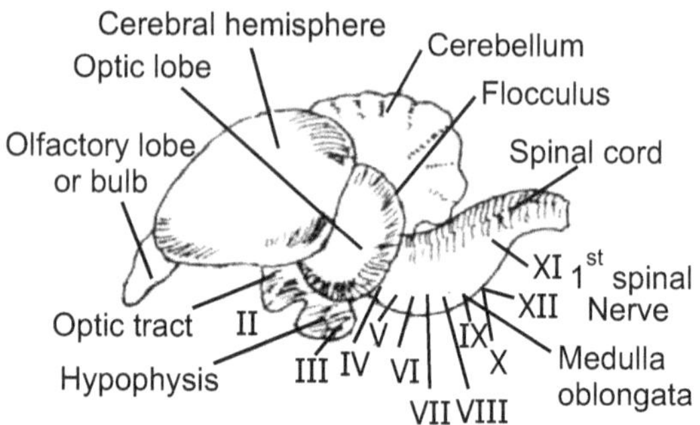

Fig. 7.6: Brain of Piegon
(a) Dorsal view, (b) Ventral view, (c) Lateral view

The brain of the pigeon is comparatively short, and broad, rounded in form and whitish in colour. It completely fills the cranium and covered by two protective meninges or membranes. The outer membrane is called dura matter made up of connective tissue and inner membrane called pia matter which is highly vascularised. The cerebrospinal fluid is filled in the narrow space between the two membanes. The brain of the pigeon also shows three divisions: namely fore brain, mid brain and hind brain.

Forebrain: The olfactory lobes are small and degenerated due to poorly developed organs of smell. The two cerebral hemispheres are very large, pyriform, convex and separated by a deep sagittal fissure.

The pallium is small and thin with some nerve cells forming a small amount of cortex as in reptiles. The cerebral hemispheres are smooth and with no convultious like mammals. The enlargement of cerebral hemispheres is due to very large and complex *corpora striata* (or hyper striatum) which are characteristic of birds, their size makes the lateral ventricles small. The cerebral hemispheres are responsible for an intelligent behaviour in birds and they control the reflex behaviour governing the lives of birds. The diencephalon is covered dorsally by the cerebral hemispheres and cerebellum. There is a single, delicate, small and median pineal body. On the ventral surface of diencephalon, two optic nerves form optic chiasma. Behind the chiasma projects a median process called infundibulum bearing a large pituitary body or hypophysis. Diencephalon relay the impulses to cerebral hemisphere and acts as integrating centre of autonomic nervous system and perception of extreme cold, heat, pain etc.

Mid brain: The mid brain is highly developed with large round optic lobes which are lateral in position because they are pressed outwards by he cerebral hemisphere. The optic lobes are connected together by transverse commissures. Optic lobes play an important role in sight.

The Hind brain: The cerebellum is large and well developed and ridged transversely. The cerebellum is large than, in mammals. It forms central part called *vermis* and two lateral lobes called *flocculi*. The large and convoluted cerebellum indicates the delicate sense of equilibrium and great power of muscular co-ordination. The medulla oblongata is concealed beneath the cerebellum and join the spinal cord. It has posterior choroid plexus. It is broader in front and tapers behind. At the point of junction of medulla oblongata and spinal cord a well marked ventral flexure as in lizards. Medulla oblongata controls all involuntary movements. From the brain 12 pairs of cranial nerves arise and leave the cranium through special holes.

7.5 Brain of Rat

Brain is the main controlling and co-ordinating centre of all body activities. It is protected in a hard, bony box called cranium. Inside the cranium brain is further enclosed by three membranes called meninges. The outer tought, protective dura matter lines the cranium, the middle delicate membrane is called *arachnoid*. The inner most membrane which closely invests the brain is the thin vascular pia matter. In between these membranes, the spaces are filled with cerebrospinal fluid. This fluid works as a lubricant and shock absorber. Thus, brain is protected from mechanical injuries or hocks. The nervous tissue of the brain is differentiated into inner grey matter which is surrounded by white matter.

The brain of the rat is divisible into three parts:

(1) Fore brain or Prosencephalon,

(2) Mid brain or Mesencephalon,

(3) Hind brain or Rhombencephalon.

Fore-brain: It is the anterior part of the brain. It consists of:

(a) Pair of olfactory lobes.

(b) Pair of large cerebral hemispheres or cerebrum, and

(c) The thalamencephalon or diencephalon.

The olfactory lobes and the cerebral hemisphere together form the telencephalon.

Olfactory lobes: These are pair of small oval structure lie at the anterior tip of the brain. They are situated on the cerebral hemisphere more towards the ventral side. The olfactory lobes are close to each other and each lobe gives out olfactory nerve to the olfactory organ. Each lobe encloses a cavity or ventricle called rhinocoel which is continuous behind with the ventricle of the cerebral hemisphere. The olfactory lobes control the sense of smell.

Cerebral hemispheres: These are pair of large oval lobes narrow and broad posteriorly. The two hemispheres lie in close contact with each other. However, the two hemispheres are separated from each other by a deep longitudinal groove called sagittal fissure. The large number of convolutions are marked on the surface of hemispheres which are correlated with different level of intelligence. The deeper and more the convolutions, the intelligence is supposed to be more. In case of rat few such convolutions are present whereas, in man many convolutions are seen. On the ventral side longitudinal fissure demarcates the hippocampal lobe on either side. The roof of the cerebral hemisphere is called pallium and its thick floor is called corpus striatum. The cerebral hemispheres are hollow, contain cavities called lateral ventricles or 1st and 2nd ventricle. The cerebral hemispheres are the seat of intelligence memory, will and thought. They also control voluntary actions.

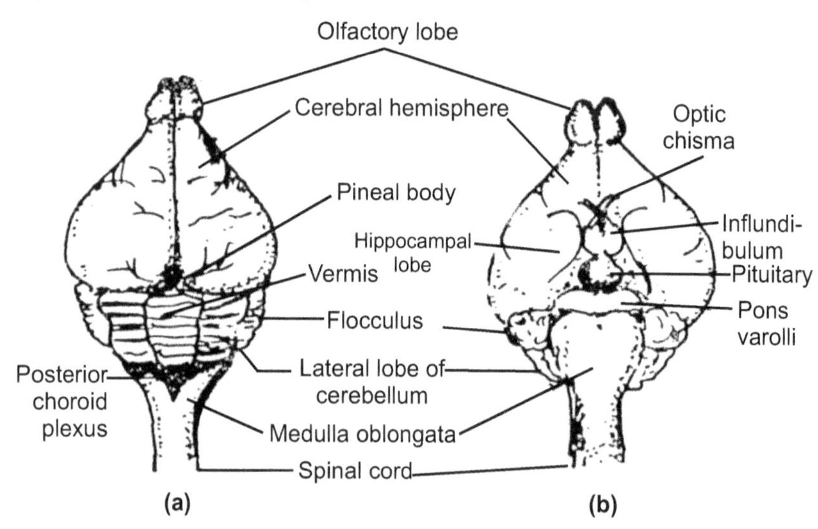

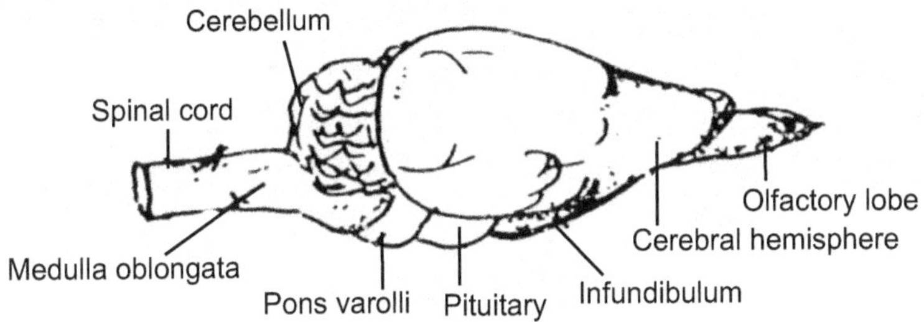

Fig. 7.7: Brain of Rat.
(a) Dorsal view, (b) Ventral view, (c) Lateral view

Diencephalon: It is a small, narrow structure and bears dorsally the stalk of the pineal body which is the functionless vestigial third eye. The lateral walls of diencephalon are thickened to form optic thalami. The roof of the diencephalon is thin, non-nervous and vascular and is called anterior choroid plexus which secretes cerebro-spinal fluid and also supplies nourishment to the anterior part of the brain.

On the ventral side of diencephalon is the optic chisma formed by the crossing of two optic nerves. From the floor of the diencephalon there arises a small outgrowth known as infundibulum whose exact function is not properly understood. Attached posteriorly to the infundibulum is pinkish rounded body, called hypophysis. The infundibulum and hypophysis together form the pituitary gland, which is the most important endocrine gland controlling all other endocrine glands. It secretes number of hormones.

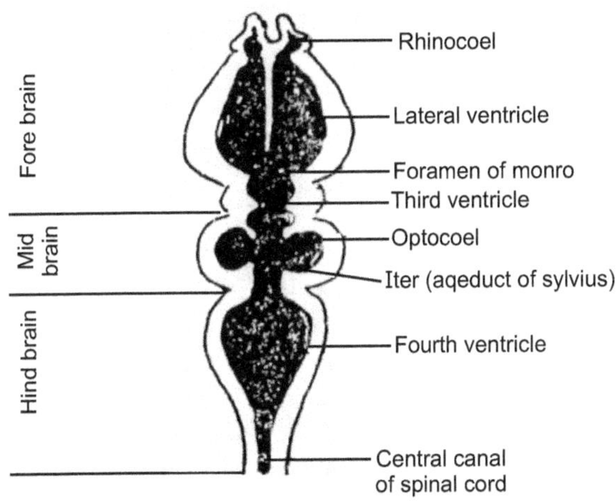

Fig. 7.8: Ventricles of brain of Rat

The diencephalon encloses the 3rd ventricle or diocoel which isnarrow and tubular. The two lateral ventricles of cerebral hemispheres open into 3rd ventricle by an opening called the foramen of Monro. The diencephalon contains centres for spontaneous movements.

Mid brain:

The mid brain or mesencephalon is thick walled region connecting the fore brain and the hind brain. The dorso-lateral surfaces of the mid brain are thickened to form the pair of optic lobes. Each optic lobe on the dorsal side is further sub-divided into two forming four optic lobes called the *corpora quadrigemina*. From the ventral surface the optic lobes give rise to two optic nerves which cross each other to form optic chiasma in front of the infundibulum. Each optic lobe encloses a cavity called optocoel. The two optocoels open into a narrow tubular cavity called iter or aquaduct of Sylvius. It connects the third ventricle with the fourth. The floor of the iter is thickened to form two strands of nervous tissue called crura cerebri through which to and fro movements of impulses are sent between the cerebrum and spinal cord. The optic lobes are concerned with the sense of sight. The posterior large lobes contain centres for auditory impulses.

Hind Brain: It consists of mainly two parts:

(1) Cerebellum, and

(2) Medulla oblongata.

Cerebellum: It is very well developed and greatly thickened. It projects forwards until it almost meets the posterior margin of the cerebrum and also extends to the lateral sides. The cerebellum is divisible into five regions and the entire surface is marked by transverse grooves. The median lobe is called the vermis, the lateral extensions are called lateral lobes. Each lateral lobe bear on outside a small outgrowth called flocculus. Cerebellum controls and co-ordinates muscular movements and thus controls equilibrium of the body.

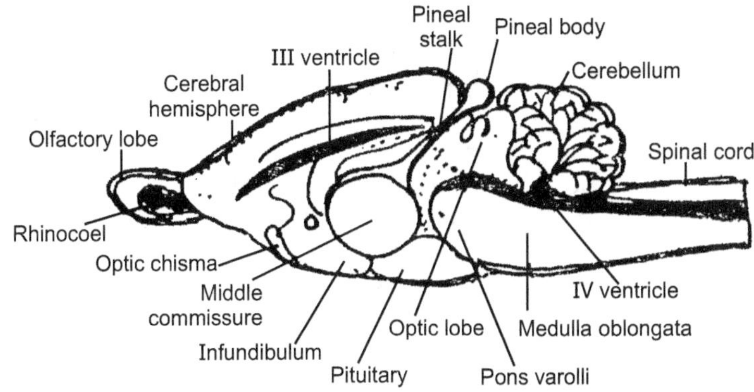

Fig. 7.9: Sagittal section of brain of Rat

Medulla oblongata: It is the posterior most part of the brain which is broad anteriorly and narrow posteriorly. The posterior end is continued further down as the spinal cord. On the ventral side of the medulla there is a stout band or nerve fibres, the pons varolii which runs transversely connecting the two floccular lobes of the cerebellum. Its ventricle is called fourth ventricle or metacoel whose thin, non-nervous vascular roof is called posterior choroid plexus. It secretes cerebro-spinal fluid and supplies nourishment to the posterior parts of the brain. Metacoel is continued behind as the central canal of the spinal cord.

The medulla oblongata contains centres controlling the involutary vital functions such as respiration, circulation, rate of heat beats, peristalsis etc. Hence, any injury to the medulla oblongata results in instantaneous death.

Functions of the Brain:

Fore-brain:
(1) Olfactory lobes are concerned with the sense of smell.
(2) Cerebral hemispheres are the centres of co-ordination and intelligence. The cerebral cortex is a seat of memory, will and intelligence.
(3) The pineal and pituitary bodies have endocrine functions.
(4) Optic thalami acts as association centres for the sense of sight.
(5) The anterior choroid plexus secretes cerebrospinal fluid and also supplies nourishment to the anterior parts of the brain.

Mid-brain:
(1) In the corpora quadrigemina, the anterior pair of lobes is concerned with the sense of sight, while the posterior pair of lobes is concerned with the relay of auditory sensation.

Hind-brain:

(1) The cerebellum controls and co-ordinates muscular movements and thus maintains the balance of body.

(2) The medulla oblongata controls the voluntary activities such as respiration, circulation, peristalsis, etc.

(3) The posterior choroid plexus secrete the cerebro-spinal fluid and supplies nourishment to the posterior part of the brain.

Points to Remember

- Brain of *Scoliodon* shows fore brain, mid brain and hind brain.
- Brain is an important organ performs different functions.
- Brain is main controlling and co-ordinating centre of all body activities.
- There are six ventricles present in the brain of frog.
- From the brain of pigeon 12 pairs of cranial nerves arise.
- Fore brain is formed by olfactory lobes, cerebral hemisphere and diencephalon.
- Mid brain is formed by optic lobes, crura cerebri in lizard.
- Hind brain of rat consists of cerebellum and medulla oblongata.
- Olfactory lobes are for sense of smell.
- Cerebral hemispheres are centres of co-ordination and intelligence.
- Pituitary body has endocrine functions.
- Optic lobes are concerned with sense of sight.
- Medulla oblongata controls the voluntary activities.
- Cerebellum maintains the balance of body.

Exercise

1. Give a comparative account of brain of vertebrates.
2. Describe the brain of lizard and compare it with that of a mammal.
3. Compare the brains of frog and rat.
4. Compare the brains of fish and frog.
5. Give an account of brain of *Scoliodon* and give the functions.
6. Describe the brain of frog.
7. Give an account of brain of lizard.
8. Describe the brain of pigeon.
9. Describe the brain of rat and give the functions.
10. Write short notes on:
    (a) *Scoliodon* brain　　　(b) Frog brain
    (c) Lizard brain　　　(d) Rat brain
    (e) Functions of brain.

✹✹✹